An Exploration of a Problem-Based
Learning Approach to the CAT Course at Tertiary
Level — An Educational Design Research Project

基于问题学习模式（PBL）的高校计算机
辅助翻译课程教学设计研究

骆雪娟◎著

U0178745

中山大學出版社
SUN YAT-SEN UNIVERSITY PRESS

·广州·

图书在版编目（CIP）数据

基于问题学习模式（PBL）的高校计算机辅助翻译课程教学设计研究 = An Exploration of a Problem-Based Learning Approach to the CAT Course at Tertiary Level — An Educational Design Research Project：英文/骆雪娟著. —广州：中山大学出版社，2021.11

ISBN 978－7－306－07196－5

Ⅰ.①基… Ⅱ.①骆… Ⅲ.①自动翻译系统—教学设计—研究—高等学校—英文 Ⅳ.①TP391.2

中国版本图书馆 CIP 数据核字（2021）第 235785 号

出 版 人：王天琪
策划编辑：熊锡源
责任编辑：熊锡源
封面设计：林绵华
责任校对：林 峥
责任技编：靳晓虹
出版发行：中山大学出版社
电　　话：编辑部 020－84111996，84111997
　　　　　发行部 020－84111998，84111981，84111160
地　　址：广州市新港西路 135 号
邮　　编：510275　　传　　真：020－84036565
网　　址：http://www.zsup.com.cn　　E-mail：zdcbs@mail.sysu.edu.cn
印 刷 者：广州一龙印刷有限公司
规　　格：880mm×1230mm　1/32　10.75 印张　470 千字
版次印次：2021 年 11 月第 1 版　　2021 年 11 月第 1 次印刷
定　　价：35.00 元

Acknowledgements

Writing a thesis of this kind inevitably involves the participation, assistance and encouragement of many people, to whom I owe a debt of gratitude.

First of all, my sincerest thanks go to Prof. HUANG Guowen, my supervisor, whose vigor and insight have always been the inspirations for me. This thesis would never have taken the present shape without his critical guidance and unflagging encouragement during the past five years. His generosity and open-mindedness as well as his rigorous and scientific spirit would benefit me far beyond the current study.

I am also deeply indebted to the School of International Studies, Sun Yat-sen University for making me who I am now. First of all, my heartfelt appreciation goes to Prof. WANG Bin. It is his recognition and assistance that has made all these efforts possible in the first place. Besides, I owe many thanks to Prof. LIN Yuyin who has always shown heartwarming readiness to help me with all the possible problems in research design and Prof. XIAO Jingyu for her unfailing care and support that has been a great comfort for me especially when I was stuck in the struggle between work and study in the past years. Special thanks are also given to Prof. XU Dongli and Mr. CHEN Youzhi for the kind support and understanding they have generously given me.

Besides, I am particularly grateful to Prof. Heather Fulford for her kindly sharing with me her survey questionnaire although we have never met each other and Prof. Florence Qi for the inspiring conversations with her during the research process.

My gratitude also goes to Prof. WANG Chuming, Prof. CHANG Chenguang, Prof. DAI Fan, Prof. MO Aiping, and Prof.

DING Jianxin for their constructive comments and insightful suggestions.

I would also like to extend my thanks to my colleagues and friends whose support and love have made such a mission impossible much easier for me. I cannot forget the books on course design and curriculum development Dr. BI Xuefei lent me when I was working on the proposal, which successfully lifted me up from the initial confusion. Equally unforgettable is Ms. ZHAO Rui who took over almost all the administrative duties from me during the past year, enabling me to fully concentrate on the thesis writing. Not to mention the generous love from many of my good friends, Ms. CHEN Shuo, Ms. XIAO Juanjuan, Ms. WENG Jinghua, Dr. ZENG Ji, and MS. MA Jichen, to name just a few. Thanks for the unwavering support you have had for me to let me know that you will be there by my side whatever happens to me. No words can express my gratitude to you, but you guys know I cannot thank you more.

Special thanks are also due to all the students who have participated in the study. I cannot list all your names here but please believe me you are and will be all living in my heart.

I need to reserve a special word of appreciation and admiration for my families, my dedicated parents, my husband who is also my mentor and friend, and my beloved and adorable son. Thank you for your understanding and tolerance. Thank you all for giving me the reasons for my life.

Abstract

The wide application of translation technology has greatly promoted the expansion of the translation industry. As a result, the ability to work with computerized tools has been globally recognized as a key skill of professional translators, a new change in translation competence that has in recent years attracted increasing attention from translator educators. The subsequent inclusion of computer-aided translation (CAT) as an independent course in translator education curriculums in higher education, therefore, is not only a sensible but also an inevitable action.

In contrast to the integration of a CAT course into translation curriculums nationwide, however, the author has researched extensively only to find that there is a dearth of systematic studies on CAT teaching in general. Further analysis suggests that those studies directed at *how* to teach CAT are much fewer than those about *what* to teach about CAT. The limited pedagogical suggestions are mostly based on personal experience, while systematic empirical exploration of educational design is still rarely seen. Meanwhile an increasing number of researchers have shown their concern with the neglect of teaching methodology in translator education at large and its undesirable consequences. The author, therefore, argues that CAT, as a new but increasingly important course in translator education with a peculiar component of technology teaching, merits more attention to its teaching methodology.

A review of the few growing studies on CAT teaching approaches suggests that ranking high among the educational aims of CAT teaching at tertiary level should be learner autonomy, problem solving, collaborative work, communication with peers, continuous learning and learning of *how* over *what*. These demands for CAT teaching reflect not only the particular built-in features of technology teaching but also the constructivist turn in higher education worldwide to which translator education is no exception. Among others, problem-based learning (PBL), widely recognized as one of the most mature constructivist learning theories, has successfully drawn attention of translator educators given its wide application across disciplines and different stages of education and the worldwide acclaims for its validity.

The envisaged alignment between CAT teaching at tertiary level and the PBL teaching philosophies, against the backdrop of the undesirably few studies on translation pedagogy, stimulated the author's interest to explore the applicability of PBL to the CAT course at tertiary level. As the first attempt to introduce PBL into CAT teaching, this study took the form of an educational design research project at Sun Yat-sen University (SYSU) with the aim to answer the following three questions: (1) How is PBL aligned with CAT teaching at tertiary level? (2) How can PBL be applied to the design of the teaching approach to the CAT course? (3) How does the PBL approach work in its preliminary implementation in the CAT course at SYSU?

Defined as *educational design research* across *exploratory* and *design & development* stages, this study opted for the *interpretivist* research paradigm and employed the research design of 3 stages using the framework of Educational Research and Development proposed by Gall *et al* (2003).

In the first stage, based on an extensive literature review, theoretical connections between PBL and CAT teaching in higher education were established from the perspectives of students, teachers and the course as integral parts of university education.

Then, in the second stage the PBL design model at course level was generated based on Barrett (2005) and then operationalized using Constructive Alignment, the constructivist instructional design framework proposed by Biggs (1996), the revised version of the Bloom Taxonomy from Anderson & Krathwohl (2001), and the theory of Zone of Proximal Development by Vygotsky (1978). The PBL approach to the CAT course of 36 academic hours and 2 credits

II

was then piloted with this design model.

The design was later field tested in the third stage by a qualitative case study in School of International Studies (SIS) at SYSU. 24 juniors from the Department of Translation and Interpreting of SIS participated in this credit-bearing project of 4 weeks. Data of different types were collected from the participating learners, the teacher and the observer via journals, questionnaires, reflective reports, assessment tasks and the observational records. Data triangulation and methodological triangulation were adopted to guarantee the reliability of the findings. QSR NVivo 10, a reputable qualitative data analysis package, was employed to facilitate data management and analysis. The data were analysed within the revised evaluation framework of Kirkpatrick (1998) using the theory-driven thematic analysis approach. Results of the analysis were reported on the three levels of reaction, learning and behaviour, based on which preliminary judgment about the feasibility of PBL in the CAT course was made.

The overall findings of the research gave us more promise than doubt about the applicability of PBL to CAT teaching in higher education.

First, at the level of reaction, the data showed a unanimously positive attitude of the participants toward the PBL approach. 16 out of the 24 participants spoke very highly of PBL supported by reasons ranging from the greatly enhanced motivation, better independent learning ability, deeper understanding and longer retention of the knowledge, increased transferability, and greater communication skills to more critical thinking as results of their learning experience with the PBL approach.

Then at the level of learning, the data of different types showed consistently that the intended learning outcomes have all been reached although to different degrees. Further probe into the self-reported data from the participants revealed to us that among the most noticeable effects of the PBL approach in the perception of the learners were those on the learners' motivation, self-directed learning, group work collaboration and problem-solving skills.

Lastly at the level of behaviour, the data, particularly those from the survey one year later, suggested that the knowledge acquired from the course was still clearly retained and the broad-sense CAT tools were indeed used often in their translation learning and practice although they did not have many opportunities to use the narrow-sense CAT tools. Yet most of the students reported the benefits in the following year they obtained from the proactive learning attitude, communication skills, perception of problems and problem solving skills that they believed to have acquired from the course with the PBL approach. That made 16 of the participants express openly their gratitude for having experienced PBL in this course.

The findings of this research provide the CAT teaching in a PBL approach at tertiary level with valuable pedagogical implications in both teaching practice and instructional design. Specifically, in terms of teaching practice, collaborative learning in small groups, problem-driven tutorial process and self-direction in learning are found to be most powerful for enhancing the teaching effects and are, therefore, highly recommended for CAT course teachers. In addition, in terms of instructional design, to design a PBL approach, the roles of teachers and students, the tutorial and learning processes and the assessment methods are advised to be well aligned in order to guarantee the teaching quality and the fair judgment of the students' learning outcomes.

内容提要

翻译技术的广泛应用促成了翻译市场的迅猛发展，能够熟练利用计算机辅助翻译工具也随之成为职业译员不可或缺的能力之一。翻译能力的这一变化逐渐引起高校教育者的关注，计算机辅助翻译（CAT）最终广泛进入高校翻译教学，成为翻译专业及英语专业翻译方向培养方案中一门独立的课程。

然而，与高校已广泛开设 CAT 课程的现状不符的是，笔者发现国内关于 CAT 教学的系统研究远远不足。对已有研究的分析表明，有关 CAT 教学的探讨多集中于教什么的问题，对于如何教却鲜有关注。少量的建议仅基于研究者的个人经验，缺乏系统的、有经验数据支撑的研究。近些年来，翻译教学中对于教学方法重要性的普遍忽视，及其可能产生的不良后果，已经引起了很多研究者的关注。鉴于此，笔者指出，CAT 作为翻译教学中越来越重要的一门课程，加之软件教学的特殊性，教学方法非常重要，急需更加深入的研究。

现有 CAT 教学方法研究文献一致指出，高校 CAT 课程应着重培养学生的自主学习能力、解决问题能力、沟通合作能力、持续学习能力、知其然更要知其所以然的态度等。这些对 CAT 教学的要求既体现了翻译技术教学自身的特点，同时也反映了世界范围内高等教育观的建构主义转向。在诸多建构主义教学理论中，基于问题的学习模式（PBL）理论，作为目前公认最为完善的建构主义学习理论之一，已经被应用于不同学科、不同教育阶段，并广受好评，近年来也开始受到翻译教学研究者的关注，引入翻译教学。

在对 PBL 的深入了解和六年 CAT 教学经验的基础上，笔者意识到 PBL 学习理论和高校 CAT 课程教学目标的高度一致性，因此试图通过将 PBL 学习模式应用于中山大学 CAT 课程，验证其在翻译技术教学中的适用性，在高校 CAT 教学方法的研究方面做出新的探索，以弥补相关研究的不足。具体来说，本研究试图通过将 PBL 应用于中山大学 CAT 课程的教学设计，回答下列三个问题：

（1）PBL 学习模式和高校 CAT 课程的一致性如何？

（2）PBL 学习模式如何应用于高校 CAT 课程的教学设计？

（3）PBL 学习模式在中山大学 CAT 课程中初步应用的效果如何？

作为跨越探索阶段和设计研发阶段的教学设计研究（educational design research），受制于此阶段研究的内在特点，本研究采纳了诠释型研究范式，参考 Gall 等人（2003）提出的教育研究和开发（Educational Research & Development）方法，设计了三个阶段的研究步骤。

第一阶段的研究在广泛的文献分析的基础上，从学生、教师及课程需求三个方面建立了 PBL 和高校 CAT 课程教学的理论一致性。

接着，第二阶段在 Barrett（2005）PBL 模型的基础上提出了本研究的 PBL 设计模型，并结合 Biggs（1996）的一致性建构理论（Constructive Alignment）、Anderson & Krathwohl（2001）的修订版布鲁姆分类理论（Bloom Taxonomy）和 Vygotsky（1978）的最近发展区（ZPD）理论，完成了针对中山大学翻译学院 CAT 课程的 PBL 教学设计。

本研究的第三阶段利用个案研究对该设计的初步实施效果加以考察。所选观察个案为中山大学翻译学院一学期 2 个学分、时长 36 课时的 CAT 课程。24 名翻译系大三学生参加了该实验课程，研究通过学生日志、问卷和总结汇报等不同形式搜集的学生自述型数据，将之和以测试为依据的客观数据以及从教师及课堂观察者角度获取的第三方数据通过三角互证的方法互相佐证，以提高数据的可信度。所有搜集所得数据均导入 QSR NVivo 10 管理，以修订后的 Kirkpatrick 教学效果评估模型（1998）为理论框架，运用理论驱动的主题分析方法（theory-driven thematic analysis）进行数据分析，获取了学生视角下 PBL 学习模式的初步应用在态度（reaction）、学习效果（learning）和行为转变（behaviour）三个层面上的评估结果。最后，在此分析结果的基础上，对 PBL 在高校 CAT 课程中的适用性做出了初步判断。

本研究的结果从总体上初步肯定了 PBL 在高校 CAT 课程中的适用性。

首先，在态度上，学生对 PBL 的引入持一致肯定意见，其中 16 名学生对 PBL 学习模式表示了高度赞赏，列举原因包括学习动机大大提高、自学能力大幅度改善、对习得的知识理解得更透彻、记忆更持久、沟通能力有所加强、更善于学以致用、思维更加缜密等。

其次，在习得效果上，无论是学生的自述型数据还是测试中的客观数据均表明，虽然完成程度有所差异，学生均比较成功地达成了课程设置的所有学习目标。根据对学生自述型数据的深入分析，PBL 最突出的效果体现在它不仅能够有效促进知识的理解和记忆，还可以明显增强学生学习动机，提高他们在自主学习、小组合作及解决问题等方面的能力。

最后，在行为转变层面，特别是时隔一年后的调查显示，学生们均表示对该课程中习得的知识记忆深刻；虽然狭义 CAT 工具的应用机会不多，但是他们在后面一年的学习和实践中广泛应用了广义 CAT 工具。通过 PBL 学习模式收获的主动学习的态度、沟通技巧、问题意识和解决问题的方法，都在他们随后一年的学习和实践中有所运用并带来了良好的效果。16 名学生在调查中表示对此次课程中的 PBL 学习经历心怀感激。

　　本次研究结果为未来高校的 CAT 课程的 PBL 教学及设计提供了很好的借鉴。具体来说，首先，在教学实践方面，本研究发现 PBL 学习模式中的小组合作的学习方式、问题驱动的教学过程以及学生自主控制的学习过程对学习效果的促进作用最明显，值得未来 CAT 课程设计者采用。其次，在 PBL 教学设计方面，教学设计者必须保证教师和学生的角色定位、教学和学习活动的过程以及考试方式之间的一致性，以确保教学质量和学习效果评估的合理性。

序

2021 年新春伊始，收到骆雪娟的微信，得知其基于博士学位论文修订而成的书稿《基于问题学习模式（PBL）的高校计算机辅助翻译课程教学设计研究》即将付梓，邀我做序。作为她的博士论文指导导师，看到她这些年的成长，真心为她高兴，所以欣然应允。

我先后于 1992 年和 1996 年在英国爱丁堡大学和威尔士大学获得应用语言学和功能语言学两个博士学位，1996 年初至 2016 年秋在中山大学外国语学院任教。1996 年开始担任功能语言学研究方向的博士生导师，2006 年起增收应用语言学方向的博士生。我是 2007 年认识骆雪娟的，当时她给我留下了很深刻的印象。她从 2009 年开始跟着我攻读应用语言学的博士学位，是我指导毕业的第 42 位博士。后来，2010 年 5 月至 2015 年 11 月，我在中山大学翻译学院兼任院长，雪娟读博期间也同时担任翻译系系主任，所以我和雪娟除了是师生关系外，还是同事关系，对她的了解也就更深入一些。

在我的印象中，雪娟真诚、开朗、好学。雪娟是 2005 年翻译学院创院时进入学院担任教学任务的第一批教师；她关心学生、热爱教学、热衷翻译，从 2007 年开始成为翻译系系主任。经过她十多年的努力，翻译系教学体系逐渐成熟，尤其是翻译技术和口译两类特色课程的建设，成绩突出。

雪娟从 2008 年开始一直致力于翻译技术类课程的开发和教

学。据我了解，早在 2005 年，翻译系就率全国之先，在本科教学计划纳入了计算机辅助翻译这一课程。但是真正落实这门全新的课程，困难重重。当时国内翻译技术方面的师资奇缺，无教学经验可循，更无现成的教材可用；少数开设此课程的高校几乎都是面向研究生，开设面向本科生的课程，近乎从零开始。当时，身为系主任的她只能自己迎难而上，承担起此门课程的设计和教学工作。我记得，2007 年她专程前往当时国内唯一拥有计算机辅助翻译专业的香港中文大学翻译系学习，并在时任翻译系系主任陈善伟教授的鼎力相助下，顺利于 2008 年在中山大学翻译学院开设此门课程。

正是出于对翻译技术领域的热爱，加上多年一线教学引发的思考和积累的经验，雪娟希望将计算机辅助翻译教学设计作为博士学位论文选题，在她向我详细介绍了相关背景知识和研究思路后，我觉得这是件很有意义的事，鼓励她对这个传统文科院系里的新兴"技术"课程，进行更为系统的反思和研究。

博士毕业后，她继续完善这门课程，与时俱进地增加了技术写作和语料库等内容。到 2020 年，已有多名学生受到这门课程的启蒙，毕业后选择了跨专业继续攻读本地化翻译或计算机辅助翻译专业。近几年，她的学生也连续在一些国家级赛事中获得好成绩，如在 2018 年第 7 届计算机辅助翻译与技术传播大赛和 2020 年的腾讯 TransSmart 第三届全国机器翻译译后编辑大赛中，她指导的本科生还获得了全国总决赛的一、二、三等奖。

如今，随着翻译专业本科和专业硕士学位的快速发展，翻译技术课程在全国已经遍地开花，北京大学、北京语言大学等更是开设了计算机辅助翻译和本地化专业。但是，针对翻译技术教学的研究多停留在翻译技术工具的评介和翻译技术课程概况的介绍上，有关"如何教"的研究仍不多见，尤其是翻译技术教学相关的实证研究几乎是一片空白。而雪娟正是进行了这个领域实证研究的尝试，因此，她的研究有以下几个亮点。

首先，在研究方法上，这是一个基于实证的教学设计研究，和同类研究相比，方法论上有所创新。她将学生自述产生的主观性评价和通过测试获得的客观数据以及从教师及课堂观察者角度获取的第三方数据通过三角互证的方法互相佐证，为研究产出的教学设计产品增强了信度。

其次，在理论方面，她在充分论证了翻译技术教学和建构主义教学理论基于问题的学习模式（PBL）的一致性后，首次尝试将 PBL 应用于高校 CAT 课程教学中，基于 Barrett（2005）PBL 模型，结合 Biggs（1996）的一致性建构理论（Constructive Alignment）、Anderson & Krathwohl（2001）的修订版布鲁姆分类理论（Bloom Taxonomy）和 Vygotsky（1978）的最近发展区（ZPD）理论，构建了 CAT 教学的 PBL 设计模型，有一定的理论创新性。

最后，雪娟的这个研究具有一定实践价值。记得年初雪娟向我报喜，她和缪君、谢桂霞两位同事合作，获得了首届翻译技术教学大赛全国一等奖，通过比赛她也发现"翻译技术类课程如何教"已成为越来越多教师一直思考的问题，并开始吸引更多研究者的关注，而这恰好是她博士学位论文的研究内容。所以我相信，在博士学位论文基础上修改的专著的出版，可以让同行分享她的研究成果，对推动国内翻译技术教学的发展有一定的参考价值。

陈善伟教授在他 2015 年的专著 *Routledge Encyclopedia of Translation Technology* 的前言中指出："翻译技术早已成为翻译实践的规范、翻译研究的重要部分、翻译教学的新范式、翻译行业的主流趋势。"国内翻译技术教学知名学者王华树博士也早在

2016 年就已指出，在"一带一路"倡议的大背景下，进一步完善翻译人才的技术能力培养和课程建设，"对于加强国家语言能力建设，促进语言服务生态系统的健康发展具有重要的战略意义"。由此可见，骆雪娟和她的翻译技术教学，仍然任重而道远。

子曰："君子不器。"（《论语·为政第二》）孔子对于弟子的要求是，君子不能像器具一样，只有一种特定的用途，而应是博学多能。无论是从政还是做学问，都应该博学且才能广泛，这样才不会像器物一样，只能作有限目的之使用。作为一个学者，努力的方向是要致力于道，要不断追求真理；"形而上者谓之道，形而下者谓之器"（《易经·系辞》）。我想，做翻译实践和做翻译研究，首先都要基于实践、以问题为导向，从实践到理论再到实践，最终是解决问题。作为一个学者，要站得高看得远。有感而发，与雪娟共勉。

是为序。

黄国文

英国爱丁堡大学博士、威尔士大学博士

华南农业大学教授、博士生导师

2021 年 2 月 25 日于华农六一区宁荫湖畔

参考文献

Anderson, L. W. & Krathwohl, D. R.（eds.）（2001）. *A Taxonomy for Learning, Teaching, and Assessing: A Revision of Bloom's Taxonomy of Educational Objectives*. New York: Longman.

Barrett, T.（2005）. Understanding problem-based learning. In Barrett, T., Mac Labhrainn, L. & Fallon, Hco.（eds.）, *Handbook of*

Enquiry & Problem-based Learning. Galway: CELT, 13-25.

Biggs, J. (1996). Enhancing teaching through constructive alignment. *Higher Education*, (32), 347-364.

Chan Sin-wai (2015). *Routledge Encyclopedia of Translation Technology*. London: Routledge.

Vygotsky, L. (1978). *Mind in Society: The Development of Higher Psychological Processes*. Cambridge, MA: Harvard University Press.

王华树 (2016), 系统论视域下的翻译技术课程建设, 《当代外语研究》第 3 期, 53-57.

Table of Contents

List of Abbreviations

AT	assessment task
BYU	Brigham Young University
CA	constructive alignment
CAT	computer-aided translation
CRA	criterion-referenced assessment
FAHQMT	fully automatic high quality machine translation
HT	human translation
ICT	information & communication technology
ILO	intended learning outcome
MAHT	machine-aided human translation
MAT	machine-aided translation
MIIS	Monterery Institute of International Studies
MT	machine translation
NRA	norm-referenced assessment
PBL	problem-based learning
R & D	research and development
SIS	School of International Studies
SOLO	structure of observed learning outcome
SYSU	Sun Yat-sen University
TM	translation memory
TMT	teaching medical translation
TT	translation technology
TTP	translator training programme
ZPD	zone of proximal development

List of Tables

List of Figures

Chapter One
Introduction

This chapter is aimed at a brief introduction to the background of the current study, against which the rationale will be explained and the research questions put forward. Explication of the research design and the significance of this study will follow before the structure of the thesis is articulated in the end.

1.1 Research background

1.1.1 Undesirable pedagogical gap in studies on CAT teaching

In recent decades, the advent of computers and the Internet has brought about revolutionary changes in the profession of translation[①]. Among them, as Gouadec (2007) depicts, one major change is the largely digitalized working environment which, together with the dramatic increase in information, has turned translation into a computer-bound and knowledge-based activity (Austermühl, 2001; Gil & Pym, 2006). Meanwhile, against the globalisation background, professional translators are required to turn out high quality translation of a greater variety of texts, in larger volumes but with a shorter time (Bowker, 2002; Stupiello, 2008). Consequently knowing how to use skillfully the great variety of computer-based tools and resources, for general and translation-specific purposes alike, has inevitably become a must for professional translators in order to keep an edge in the ever fiercer competition on

---

① Translation in this thesis is exclusive of interpreting and refers only to the activity of providing written language-mediating services.

the translation market by promoting their efficiency at no cost of the quality (Gouadec, 2007). Confronted with the dramatically changed demands for translation services, the common problem facing the translator training programmes (TTPs) both at home and abroad nowadays is how to accommodate adequately and flexibly the market needs and the rapidly changing translator profile in the professional world (Chai, 2010).

Aware of the increasing importance of computer-based tools for professional translators in the workspace, many researchers of translation teaching have long since addressed this issue in both teaching practice and theoretical studies. The call for more attention from translator educators of higher education to the necessity to integrate translation tools training into the curriculum of TTPs started as early as in the 1990s in the west (e.g. Chan, 2007b; Kenny, 1999; Schäler, 1998). For example, early in 1996, Kingscott (1996) has warned that "unless technology-related issues are integrated into translator-training programmes, there is a real danger that the university teaching of translation may become so remote from practice that it will be marginalized and consequently be widely perceived as irrelevant to the translation task" (p. 295 cited from Bowker, 2002, pp. 13-14). Almost 10 years later some Chinese researchers sensitive to the market changes joined in this effort (Lv & Mu, 2007; Xu, 2006; Yuan, 2005) which paid off when the course of Computer-aided Translation (CAT) was approved by the academic authorities as an elective at both undergraduate and postgraduate levels of TTPs in China (Zhong, 2007; Zhong, 2011). Moreover, the number of TTPs which have offered CAT training is recently increasing quickly. As an informal survey the author conducted among 30 universities with TTPs shows, 22 of them have offered CAT training in the form of a course or a series of courses, among which 9 open it to both graduate and undergraduate students. The content of the courses has wide coverage ranging from narrow-sense CAT technologies (e.g. Trados, MemoQ) to translation project management and even knowledge capitalization skills. Yet CAT was adopted almost unanimously as the course name very probably because this term has been extensively used by software developers, professional translators, and in recent years, more and more by translator educators as well as translation researchers in China.

Now it has been widely acknowledged that professional translators today are increasingly dependent on a wide variety of tools and resources. Advantages of integrating CAT teaching into the translator training curriculum range from meeting the market needs to fostering the translator's self-awareness, self-concept, and a sense of the profession which lead some researchers to even argue for a central standing of CAT teaching in the translation curriculum.

Meanwhile, the importance of the ability to use CAT tools is also reflected in theoretical studies of translation teaching. Studies of translation competence, a prerequisite for effective translator education, has been gaining greater attention and becomes a thriving new area. Most of the models proposed of translation competence are multicomponential. Skillful use of the tools and resources has been widely believed to be so important that it secures for itself an essential part in most models of translation competence that distinguishes the professional from the layman (though still subject to empirical verification), conceptualized in various terms denoting an instrumental sub-competence (EMT Expert Group, 2009; Göpferich, 2009; PACTE, 2005, 2009, 2011; Qian, 2011; Wen, 2005).

However, regardless of the wide recognition of the importance of CAT training in translator education, a literature review presents us an undesirable lopsided view: a vast majority of the studies are devoted to introducing the newly developed CAT technologies and their application in translation practice; much less are concerned with the teaching of the technologies (e.g. Luo, 2010; Qian, 2009; Samson, 2010; Xu, 2010a). This is evidenced by the figure contrast: a search on CNKI with the key word "CAT" produces 242 results while with an additional key word of "teaching", the figure slumped to only 14. This phenomenon is not peculiar to China.

Worse still, a deeper probe into the prior literature shows that among the studies concerning the teaching of CAT technology worldwide, most of the research available was devoted to what to teach about CAT while studies focused on the teaching method of CAT are unreasonably few among which, worse still, empirical investigations are in severe scarcity, with most of them drawing on personal experience or knowledge only.

The ignorance of teaching methodology for CAT courses is undesirable seeing, for one thing, the fact that no teaching is expected to make success without attention to how it is executed, and for the other, the educational shift in paradigm is taking place in the broader context of higher education from teacher-centredness to learner- and learning-centredness based on the verified cognitive and psychological learning theories.

Kiraly (1995) has long before expressed his concern over the ignorance of the importance of systematic approaches to the teaching of translation skills at large. The gap between translation teaching practice and translation studies as well as pedagogical theories, as Kiraly (ibid.) noted, would undermine the chances of producing quality translators. Now CAT training is in a similar situation. In fact, as translation has only recently arisen as a new discipline in China, translation pedagogy in TTPs is a new area to be explored at large. Many Chinese scholars have begun to call for more studies in this regard (e.g. Miao & Liu, 2010; Wen, 2005; Wen & Li, 2010; Zheng & Mu, 2007).

6 years of teaching CAT courses in the university makes the author believe that the new demands from the translation profession and the intrinsic features of technology teaching combine to justify the need for more systematic research on innovating teaching approaches to CAT training in higher education.

1.1.2 PBL as a promising alternative teaching approach

Extensive research on the new demands for higher education in general and CAT teaching in particular combined with that on the advancements in learning theories reveals to the author problem-based learning (PBL) as the most promising alternative teaching approach to the CAT course.

On the one hand, having been recognized as among the core competence of modern professional translators, CAT training in higher education in China almost unanimously takes place in the form of a course on TTP curriculum. An innovative teaching approach taken by this course will have to be able to address both the external demands from the professional world and the internal needs as an integral part of higher education. As reported in the guidelines for the national curriculum design of TTPs in China, Zhong (2011) stressed the importance of adopting learner-centred approach to classroom teaching with an aim to enhance students' learning autonomy, preparing them to be life-time learners, which actually represents an international trend in higher education. In light of this the CAT course in higher education nowadays has to go beyond the traditional role of merely imparting domain-specific knowledge to students. Research shows that it has to be taught in such a way that meets the demands for:

(1) mastery of computer-based tools and resources to ensure efficiency and quality of translation and facilitate teamwork in the translation process;
(2) awareness of and competence in translation as a social communicative task;
(3) willingness and ability to work in groups as well as individually;
(4) positive attitude towards challenges and adequate problem solving skills;
(5) self-motivated continuous or life-time learner.

On the other hand, in the constructivist turn in higher education, PBL, as one of the best exemplars of constructivist pedagogies, emerged as a learning approach characterized by using ill-structured authentic problems as stimuli of the learning process self-directed with tutors acting as facilitators and students learning in collaboration and group work (Barrows, 1996; Graaff & Kolmos, 2003; Savery & Duffy, 1995; Taylor & Maflin, 2008). It has been widely welcomed across disciplines (e.g. medicine, business, science) in different stages of education (e.g. K-12 and higher education), and in recent years warmly ushered into translator education (e.g. Huang & Wang, 2012a, 2012b; Inoue, 2005; Sánchez-Gijón, Aguilar-Amat, Mesa-Lao & Solé, 2009; Stewart, Orbán & Kornelius, 2010).

The fact that it has won world-wide acclaims and is regarded by Savery and Duffy (1995) as "one of the best exemplars of an instructional model with a clear link between the theoretical principles of constructivism, the practice of instructional design, and the practice of teaching" (p. 31) assures the author of not only its great potentials but also its readiness for application in a broader context.

In sum, envisaging PBL's great potential for enhancing CAT teaching quality, the author aimed to address the general neglect of research on CAT teaching methodology in higher education by undertaking the exploration of a PBL approach to CAT teaching at course level.

1.2   Research questions and research design

In brief, this study was aimed at innovating the teaching methodology for the CAT course in undergraduate-level TTPs under the guidance of the constructivist PBL theory. As the first attempt in this regard, the author pioneered an educational design research project at Sun Yat-sen University (SYSU) and formulated the following research questions:

(1) How is PBL aligned with CAT teaching at the tertiary level?

(2) How can PBL be applied to the design of the teaching approach to the CAT course?

(3) How does the PBL approach work in its preliminary implementation in the CAT course at SYSU?

These research questions situated this study across exploratory and design & development stages (Institute of Educational Sciences *et al.*, 2013). Consistent with the underpinning constructivist philosophy of the PBL approach, the study opted for the interpretive research paradigm assuming that researchers construct their understanding of the research objects by experiencing them. By implication qualitative methodology is preferred to quantitative one and the actors experiencing the research objects are the key to the researchers' understanding of them.

First, Educational Research and Development (R&D) proposed by M. Gall, J. Gall and Borg (2003) was adopted as a framework for the research design. Five steps, regrouped into three stages, namely Building Theoretical Connections, Pilot Design, and Preliminary Field Test, were taken to achieve the four research objectives:

(1) to build theoretical connections between PBL to CAT teaching in higher education;
(2) to pilot a design of a PBL approach to CAT teaching practice;
(3) to carry out a preliminary field test of the design;
(4) to inform future formal application of the PBL approach to CAT teaching with the empirical assessment of the initial implementation.

More specifically, the definition of PBL was clarified and its alignment with CAT teaching in higher education was built based upon an extensive and systematic review of literature in the first stage.

Following it a CAT course using PBL approach was designed under guidance the prototype PBL model revised from Barret (2005) using Constructive Alignment (CA), the constructivist instructional design method proposed by Biggs (1996). Moreover the revised version of the Bloom Taxonomy from Anderson and Krathwohl (2001) was adopted in phrasing the intended learning outcomes (ILOs) and the theory of Zone of Proximal Development (ZPD) by Vygotsky (1978) was used to guide the teaching content sequencing.

The design was then implemented and field tested using a case study on the CAT course offered to juniors in Department of Translation and Interpreting in School of International Studies (SIS), SYSU. 24 leaners participated in the credit-bearing project of 36 academic hours that took place in the first term of Academic Year 2013 starting from August 12. Method triangulation and data triangulation were adopted during the data collection phase to improve the trustworthiness of the results of the test. More specifically, data of different types were collected from different sources, which include self-reported data from the participating learners and the teacher via journals, questionnaires and reflective reports, and factual data from the assessment tasks and the observational records made by the classroom observer. Given the huge amount of data produced during the qualitative research, QSR NVivo 10, a reputable qualitative data analysis package, was employed to facilitate the management and analysis of the data. The data were then analysed within the revised version of the reputable evaluation framework by Kirkpatrick (1998), using the theory-driven thematic analysis approach. The preliminary teaching effects were then constructed from the perspective of participants and reported on the three levels of reaction, learning and behaviour, from which pedagogical implicatons and further research directions were also identified.

1.3 Significance of this study

This research pioneered a pedagogical innovation in CAT teaching in higher education by leveraging recent advances in learning theories, educational design research, translation technology (TT) and translation studies. It is expected to be of significance in at least the following three aspects:

First of all, theoretically speaking, this successful attempt at applying PBL to a new discipline does not only contribute to expanding the theory's applicability, but also lends more supporting evidence for its validity in a wider context.

Secondly, practically speaking, by introducing PBL into CAT teaching in higher education, it was intended to make a contribution to improving the compatibility of the current teaching practice in CAT courses with the present market needs. It is hoped that the PBL intervention for the CAT course could help address the new challenges from the translation market to CAT teaching and enhance its quality. The author also expects her design to shed some light on the directions for pedagogical innovation in translator education in general against the constructivist turn in higher education internationally.

Lastly, methodologically speaking, the research design of this study is hoped to provide researchers of common interest with valuable experience in initial-stage theory-informed and empirically based pedagogical design for CAT courses in particular and courses in TTPs in general. Situated in *educational design research* at its early stage, such preliminary investigation at the beginning of the design research cycle was usually wrongly neglected and mistaken as "scientifically insignificant" (Akker, Gravemeijer, McKenney & Nieveen, 2006, p. 147) as viewed from the dominating positivist paradigm. Now, however, it has been recognized as crucial effort without which the circle could not have got started and become an "emerging trend" (ibid.). Yet as a new type of research which is still "characterized by a proliferation of terminology and a lack of consensus on definitions" (ibid., p. 47), there is no sufficient antecedents to follow. This study, therefore, made some bold effort with the methodological design which is expected to be of value for colleagues with similar needs or interests.

1. 4   Structure of the thesis

This thesis consists of seven chapters.

Chapter One provides an overview of this study, including the research background, the rationale for the current research topic, and the research questions addressed followed by the explication of the research design and its significance.

Chapter Two is devoted to an in-depth literature review of the history of CAT teaching in higher education and the status quo of studies on it worldwide. Analysis ensues to show how the choice of the research topic is justified and how the research questions arise.

In Chapter Three the research design is explicated in light of the research objectives, with detailed introduction to the research paradigm of choice, the research process and the methodology adopted for each stage of the study.

Chapter Four follows, as a report on the result of the first stage of the study, to clarify the concept and theoretical foundations of PBL on the basis of which its alignment with CAT teaching in higher education is established.

Then Chapter Five moves on to the second stage, with an attempt at a pilot design of a PBL approach to the CAT course in the context of SIS, SYSU which then is field tested with the findings displayed and anlysed in Chapter Six.

The findings are then discussed in response to the research questions, with pedagogical implications drawn and future research directions pointed out in Chapter Seven. Limitations of this study are explained before conclusions are made in this closing chapter of the thesis.

Chapter Two
CAT Teaching in Higher Education: An Overview

This chapter is dedicated to a tour of the studies concerning CAT teaching to provide an overview of the background of the current study, leading justifiably to its research focus. Firstly, given the vagueness of CAT as a term, an overview of the concept evolution will be provided in which a set of related and synonymous terms are clarified to prepare for a consistent use of terminology in this study. Following it is a review of studies on CAT and its role in translation competence which is a prerequisite to systematic and effective CAT teaching. A sketch of CAT teaching in translator education will ensue with a closer look at its history and the status quo where an undesirable pedagogical gap is disclosed. PBL, a promising innovative instructional strategy for professional development, will then be introduced and justified as a worthwhile candidate pedagogy for CAT teaching in higher education, from which the research questions arise.

2. 1　A sketch of CAT: concept and development

In this section, the inception and development of CAT will be sketched through a comparison of two sets of concepts: (1) CAT and Machine Translation (MT), and (2) CAT, translation technologies (TTs) and information & communications technologies (ICTs).

2. 1. 1　MT and CAT

It is based on the following two considerations that the author decides to look at MT together with CAT: for one thing, these two concepts are so interwoven with each other in the history of TT that it is difficult to understand thoroughly the conception and development of CAT without knowledge of MT; for the other, with a shorter history and less familiarity to the public, CAT is very often confused with MT by the general public. Therefore, a complete picture of these two concepts is given here to facilitate understanding of CAT.

2. 1. 1. 1 A brief history of MT

MT technology has gone through many ups and downs to become what it is today.

The computer came in reality to be involved in translation when the idea was brought forth by Weaver between 1947 and 1949 about using the new digital computers for translation. His *Memorandum on Translation* is widely believed to mark the appearance of the field of MT. Weaver's proposal was not indisputably applauded yet supporters of his began to take action very soon. MT has henceforth turned into a practical cause rather than a mere fantasy in modern science fiction. People aspired fervently to fully automatic MT, known as Fully Automatic High Quality Machine Translation (FAHQMT)[1], which was expected to replace human translators altogether someday. In 1951, Yehoshua Bar-Hillel was appointed to the first full-time research post in this field by MIT. Just one year later MIT hosted the first conference on MT. A Georgetown MT research team followed closely and demonstrated its system IBM Mark II to the public in 1954. A number of MT research programs were started hence and attracted increasingly large grants and great attention from governments of many countries, among which are the US, Japan, the USSR, Great Britain, and Canada. Researchers' continued zest for this field brought about the Association for MT and Computational Linguistics in the US in 1962.

The first turning point in the history of MT came in the year

---

[1] FAHQT means translation that is performed wholly by the computer, totally independent from human translators, but attains "high quality", which is a goal that was and is seen as utopian and will be so even in the foreseeable future if no controls or restraints are applied to the language in the texts.

1964 when the Automatic Language Processing Advisory Committee (ALPAC) was formed by the National Academy of Sciences of the US to study if the government's investment in this field was worthwhile. The infamous ALPAC report got published in 1966, concluding that with lower speed, less accuracy and higher cost than human translation, MT could "not sensibly" contribute to "any foreseeable improvement in practical translation" (ALPAC, 1966, p. 38). This conclusion, though "condemned as biased and short-sighted" (Hutchins, 2005), severely shattered the dream of FAHQMT and led to a period of stagnation in MT research that lasted for around 10 years.

Funding for MT was then greatly reduced and MT research declined quickly across the world, though it never came to a real stop, especially outside the US. Several MT systems managed to find their way to use, such as the Systran system employed by the USAF in 1970 and the Meteo system used for translating weather reports in Canada since 1976. But the nadir of MT did not end until early 1980s when increased computational power paved the way for statistical models for MT which could produce a better result in translation. The improved quality together with the changed market needs over time restored people's faith in MT. The availability of microcomputers made possible the wider application of MT. Various MT companies were hence launched, including Trados which was among the first batch of commercial translation memory (TM) technologies and came all the way to be a leading brand to date.

The end of 1990s witnessed the second turning point of MT technology when several major innovations took place, namely statistical and corpus-based approaches to MT and the start of research on speech translation. More importantly MT activity went beyond mere research toward practical applications (Hutchins, 2005). Ever since then, MT has developed rapidly to meet an increasing variety of needs ranging from individual Internet users' real-time online service to customized corporation localization facilities.

In prospect, MT is expected to be a powerful aid for translation with further research on statistical & example-based models and development of speech translation for specific domains in integration with other language technologies (ibid.).

2. 1. 1. 2　A brief history of CAT

Computer-aided translation as a concept might begin to draw serious attention since the ALPAC report. Its conclusion, while sending MT to its doom (though provisional), directed people to a new yet more feasible desire for "machine-aided translation" (ALPAC, 1966, p. 39; Somers, 2003, pp. 4-8). The two important CAT systems in operation mentioned by ALPAC were *the Federal Armed Forces Translation Agency, Mannheim, Germany*, and *the Terminological Bureau of the European Coal and Steel Community, Luxembourg*, both of which provided glossary look-up for translators and were assessed as of value in one way or another. It was believed that aids like this could be much more economically effective as a support for translation production than any of the current MT systems. It was probably from this time on that CAT[①] (often alternated with Machine-aided Translation, i.e. MAT or Machine-aided Human Translation, i.e. MAHT) as a term came into use in contrast to MT and Human Translation (HT) to indicate a different degree of human interference in the work of translation.

In telling the story of CAT, the first name that has to be mentioned is Yehoshua Bar-Hillel who had pointed out the non-feasibility of FAHQT as the very goal of the whole MT research six years before the ALPAC report. It was as early as in his review in 1960 that he argued, against the overwhelming frenzy of MT as a replacement of human translators, for a less ambitious and more practical prospect of "developing electronic machinery as aids in translation" (Bar-Hillel, 1960, p. 118) and promoted the usefulness of human translation to a higher level. It is according to this thought that Martin Kay (1997), a second must-mention figure in the history

---

① CAT and MAT are used interchangeably by some authors, but CAT is often used in Translation Studies (TS) and the localization industry while the software community prefers to call it MAT (Quah, 2008, p. 6). Similarly "aided" is also replaceable by "assisted". For the sake of consistency, this thesis will adopt "computer-aided translation", shortened as "CAT" throughout.

of CAT, proposed in 1980 a more practical idea about cooperative man-machine systems, called "translator's amanuensis". This Translator's Amanuensis was actually a network of terminals linked to a mainframe computer that would provide various levels of facilities to satisfy the translator's needs. The facilities he envisaged included mainly a special text processing editor, some add-ons such as a dictionary, a database with present and past material stored and adapted to actual use, and an interactive machine translation system. Such an idea of integrated translation-oriented application is later more often addressed as workstation or workbench. Hutchins (1998), the third prolific researcher in the MT field, explored in depth the origins of the translator's workstation and based on his survey made the most comprehensive summary of the incorporated facilities:

> ...multilingual word processing, OCR scanning, electronic transmission and receipt of documents, spelling and grammar checkers (and perhaps style checkers and drafting aids), publication software, terminology management systems, software for concordance and text analysis, access to local or remote termbanks (or other resources), TMs (for access to individual or corporate translations), and access to automatic translation software. (p. 289)

Since the 1990s CAT as a term was more widely adopted to refer to these highly-specialized translation systems which integrate various tools and functions, such as SDL Trados Studio by SDL International or Déjà Vu by Atril. The latest study on the workstation is Somers (2003, pp. 14-29) who includes as part of the translator's workstation the word processor with functionalities such as a word-count, a spell-checker or a thesaurus, lexical resources such as on-line dictionaries, term banks and encyclopedias, as well as a TM and other corpus-based resources such as alignment functions and concordance search. Among these features, TM and term banks are the most essential components.

Nowadays in the globalized economy, an ever increasing amount of translation needs to be done within shorter turnaround times in multiple languages. Group work is then becoming a must and translation management, therefore, emerges as a great challenge. In response to the new trend, many CAT products, such as SDL Trados, Across, XTM, have begun to integrate the new functionality of project management and workflow control.

2. 1. 1. 3　The differences between MT and CAT

In this section the differences between MT and CAT are to be made clearer. Two factors are often mentioned in distinguishing these two technologies. One is the involvement of humans. The most widely known and cited classification of translation in relation to technology is that made by Hutchins and Somers (1992, p. 148) as shown below in Figure 2. 1.

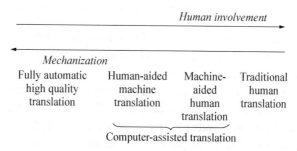

Figure 2. 1　Hutchins and Somers' scale for measuring translation automation（1992）

This classification represents the early definitions of MT that include under this term only those automatic systems with no human involvement (e.g. Sager, 1994; Hutchins, 2000). However, there are other definitions of MT which do not exclude human involvement. For instance, Arnold, Balkan, Humphreys, Meijer and Sadler (1994) indicated some human intervention in their definition of MT as "the attempt to automate all or *part of* the process of translating from one human language to another" (p. 1). Later the boundary between MT and CAT in terms of human involvement became "murky" (Balkan 1992, p. 408). MT has since then been used more often to refer to both the fully automatic systems and those with human involvement (e.g. Archer, 2002; Somers, 2003).

The second factor that separates MT from CAT is the different focuses of translation, i.e. who does the essential part of the translation. In an MT system, there is an MT engine or language parser which does most of the translation by analysing the structure of source text, breaking this structure down into elements and then recomposing a term of the same structure in the target language while human assistance may come in only at the text-preparation stage or the output stage, normally known as "pre-editing" and "post-editing" (Quah, 2008, p. 11). On the other hand, the translation is still carried out by human translators in CAT systems. In other words, human translators are in full charge of the translation process using CAT products (composing a variety of tools integrated into a workstation as described in Section 2.1.1.2) as assistance only to improve their efficiency, quality and facilitate the cooperation in teamwork. For example, translators may make a consistent use of terminology with the help of the terminology management tool. All human translations can be stored in a repository called a TM which will be updated every time a new product is translated on it so that the translated text and terminology used can be consistent within all versions of a product.

In fact, however, the boundary between these two technologies can hardly be clear-cut and has been even further obscured recently when CAT tools are turning more varied and multifunctional. The argument over MT and CAT, especially about which subsuming which has lasted till now with no conclusive answer (Chan, 2007b). Especially arguable now is the categorisation of human-aided machine translation and machine-aided human translation under one umbrella term of "computer-assisted translation" (cf. Chellamuthu, 2002; Quah, 2008, p. 8; Tong, 1994, p. 4, 730).

The current study, however, sees no point keeping on about this argument, given that human involvement in translation process is more or less necessary since FAHQT is not to be realized in the foreseeable future and that nowadays with the improved performance of the MT technology, most specialized translation systems tend to integrate both an MT and a TM① components, such as SDL Trados and Google Translator Toolkit. Consequently, for pedagogical purposes this study tends to adopt a broad definition of CAT which subsumes both HAMT and MAHT as defined by Hutchins & Somers (1992).

2.1.2 CAT, TTs and Information & Communications Technologies (ICTs)

A new trend in the denotation of CAT is that with the rapid advancement in information technologies, some authors began to advocate a wider definition of CAT to include any computerized tools and resources, translation-dedicated or not, that are of help throughout the whole process of not only "translation as a linguistic and cultural process" but also "translation as a business" (Austermühl, 2001, p. 11). CAT technologies under this definition may range from document creation to financial management tools and even such hardware as computers and scanners (Fulford & Granell-Zafra, 2005; Locke, 2005; Samson, 2010; Xu, 2010b). New terms begin to emerge when discussing translators' computerised work environment. Hence, a few comments will be made below to further clarify the definition of this study.

As depicted in Section 2.1.1.2, developed out of the conception of a workstation for translators, CAT is coined to refer to a variety of electronic tools which are of help in the translation process. Due to the fact that CAT as a term remains imprecise and loosely defined, it is found in literature that when discussing the range of tools and resources translators use in translation process, researchers also use *broad-sense CAT* (Xu, Guo & Guo, 2007), *Translation Technologies* (e.g. Alcina, 2008; Wang & Lu 2008), *Electronic Tools for Translators* (e.g. Austermühl, 2001) or *Information and Communications Technologies* (ICTs) (e.g. Heather

---

① TM is the shortened form of Translation Memory which is the very essential component of any specialized CAT systems/workstations. Sometimes it is used by some researchers (especially of early times) alternately with CAT.

& Joaquín, 2005) as an alternative. Among them, TTs and ICTs seem to be gaining increasing favour in recent publications (e.g. Fulford, 2005; Gouadec, 2007).

Table 2.1   Freelance Translators: Activities and ICT Support
(Fulford & Granell-Zafra, 2005)

Freelance Translators: Activities and ICT Support	
Activity	ICT Support
Document production *e.g. creating and formatting target texts; overtyping sources texts with target texts*	Word processing software (e.g. MS Word, WordPerfect); Graphical / presentation software (e.g. MS PowerPoint); Web publishing software (e.g. MS FrontPage, Dreamweaver); Desktop publishing software (e.g. QuarkXPress, PageMaker)
Information search & retrieval *e.g. locating background and reference materials; locating client company information; identifying terminology; locating definitions of terms; finding examples of terminology usage; managing personal terminology collections*	Internet search engine (e.g. Google, Altavista); Electronic encyclopedias / reference work (e.g. Encyclopedias Britannica, Encarta); Terminology databank (e.g. Euro-DicAutom, CILF); Text corpus / document archive (e.g. British National Corpus, New Scientist Archive); Electronic library (e.g. The British Library, Biblioteca Nacional de Espana); Electronic dictionary and / or glossary (e.g. yourDictionary.com, Lexicool); Database software (e.g. MS Access, FileMaker); Terminology management software (e.g. MultiTerm, Lingos)
Translation creation *formulating translation*	Translation memory (e.g. Trados, Deja Vu, SDLX, Transit); Machine translation (e.g. Reverso Pro, Systran)

Continued

Freelance Translators: Activities and ICT Support	
Communication *e.g. liaising with clients; networking with colleagues*	Electronic mail (e.g. Webmail, MS Outlook, Thunderbird); Electronic mailing lists (e.g. LANTRA-L, The LINGUIST List); Online discussion groups (e.g. Proz. com, TranslatorsCafé. com)
Marketing & work procurement *e.g. promoting translation services; searching for clients; bidding for translation contracts*	Having own web site; Online marketplaces (e.g. Foreign-word. biz, Proz. com)
Business management *e.g. client & contact data management; contract quotations; billing / invoicing; financial management*	Database software (e.g. MS Access, FileMaker) Spreadsheet software (e.g. MS Excel, Lotus 1-2-3); Accounting / bookkeeping package (e.g. Sage, QuickBooks)

Xu *et al.* (2007) proposed a broad definition of CAT software which includes six types of tools. They are word processors, OCRs, electronic dictionaries, electronic encyclopedias, search engines, and desktop search tools. In addition to the software proposed by Xu *et al.* (ibid.) under *broad-sense CAT*, Qian (2011) in her definition included even such hardware as the computer, the scanner, and the recorder. So far the most comprehensive definition of CAT tools may be provided by Gouadec (2007, pp. 264-280). Gouadec (ibid.) included in CAT tools both dedicated translation aids and general information technologies that are of help for translators. He further divides them up into the following four categories: (1) basic hardware equipment, (2) text processing and desktop publishing, (3) web site editing/creation software, and (4) translator tools or tools for translation (e.g. search engines, terminology and phraseology management software, TM management systems, voice recognition software, translation management software, and machine translation systems requiring human intervention).

Wang and Lu (2008) adopted the term of *TTs* and classified

them into four types based on previous research, namely CAT tools, MT tools, general tools and electronic resources. Alcina (2008) also opted for the term *TTs*. Yet her scope is wider than Wang and Lu (2008) to include the five blocs as follows: (1) the translator's computer equipment, (2) communication and documentation tools, (3) text edition and desktop publishing, (4) language tools and resources, and (5) translation tools.

Different from the above-mentioned researchers, Austermühl (2001, p. 11) was probably the first researcher to propose openly a process-oriented approach to CAT. By process, he refers to not only translation as a business process but also a linguistic and cultural process. But in later elaboration he still gave much of his attention to tools supporting linguistic and cultural transfer and left the business process almost untouched. Along this line, Locke (2005) took a process-oriented look at the translation workflow of a translator and broadened translation activities to include budgeting and pricing. In their study, Fulford and Granell-Zafra (2005) went further to bring into translation activities marketing, work procurement, communication/client liaison, bookkeeping/financial management, and even billing/invoicing. Thus in later survey of the ICT uptake of freelancers in the UK they focused not only on such tools as word processor, TM, terminology management which are used to support core translation activities, but also those adopted to communicate with clients (e.g. emails), search for clients (e.g. online market places) and manage their own business (e.g. database software, financial management tools). This might be the broadest coverage ever of computerized tools for translation.

2. 1. 3   A summary

As depicted in this chapter, the relationship between computer and translation began with MT and, in spite of some setbacks in history, has since the 1980s developed steadily till today when the computer is unanimously acknowledged as an integral part of the infrastructure needed by translators. Translation-related technologies are gaining momentum in recent years, spurred by the soaring demand for translation service on the one hand and the rapid technological advancement in computation on the other. Computerized aids for translators have, therefore, increased considerably not only in number but also in variety. New terms have hence been coined one after another to address this change. Those terms which have gained enough recognition include MT, CAT, broad-sense CAT, TTs, as well as ICTs. Although the coverage of tools under such terms varies more or less between even individual users, what seems to be commonly accepted is that a process-oriented view of the computerized tools for translators is increasingly favoured by researchers.

Yet when it comes to translator education, it is not always the more the better. Admittedly, as Samson (2010) argued, since prospect graduates may possibly take different routes (e.g. working as in-house or freelance translators) or work on different positions (e.g. as a terminology manager, a localization translator, or a quality control specialist), to give them opportunities to be conversant with a variety of general and specific computer skills seems to be more desirable from an educator's point of view. Yet every translator education programme should take into consideration all the possible constraints (e.g. time, staff, students' background) before deciding on the scope of the CAT tools to be taught so as to ensure the feasibility.

To sum up, this study tends to follow Austermühl's (2001) lead, proposes a process-oriented definition of CAT with a wide coverage of the computerized translation aids and suggests for translator educators a flexible teaching design with consideration of the feasibility in specific contexts.

2.2 CAT and translation competence

The mushrooming TTPs around the world have led to a booming interest in the studies of translation teaching. Among them, translation competence, as a prerequisite for effective translation teaching, has received ever greater attention. In this section the author will try to collect the published endeavours to understand translation competence both theoretically and empirically and display the place of CAT in them.

2. 2. 1   Theoretical efforts

Researchers have made various attempts at modeling translation competence (see Wang & Wang, 2008 for an overview). In the past forty-odd years, the discussion about translation competence has centred on subjects from purely linguistic definitions through cognitive and constructivist representations to professionally focused modelling in recent years. The latest development is that there is increasing consent among researchers that translation competence is composed of a number of identifiable and interdependent subcomponents (Colina, 2009; Márta, 2008; Miao, 2007). However, what subcomponents should be included varies from one scholar to another and remains a highly controversial subject.

Regardless of the disputes over the possible components of translation competence, in most of the currently dominant multi-componential models of translation competence, especially those coming out after the year 2000, *translators' mastery of computerized tools and resources* is almost a must component, termed differently as Instrumental Sub-competence in PACTE (2003), IT Skills in Wen (2005, p. 64) and Wen and Li (2010), Instrumental Competence (*gongju nengli*) in Wang and Wang (2008), Technological Competence in EMT Expert Group (2009, p. 7), and Tools and Research Competence in Göpferich (2009) and so on.

Unfortunately many of the afore-mentioned concepts are not specified enough to be readily adopted in the educational context. For example, Göpferich (2009) gave no definition of his *Tools and Research Competence*, only assuming that it is among the competences which separate professional translators from bilingual persons without training in translation. Wen and Li (2010) pointed out openly that what should be taught of IT Skills is still subject to further discussion.

Some other researchers have attempted to give definitions to these concepts, though very briefly. Among them, Wang and Wang (2008) included under their *Instrumental Competence* the ability to use bilingual dictionaries, collocation dictionaries, parallel texts, translation corpora, online resources and the ability to communicate with experts, although they did not explain further why they proposed these tools and resources rather than others and neither did they distinguish electronic tools from the traditional ones.

EMT Expert Group (2009), in response to the changed challenges from the market, provided a reference framework of competences necessary for the translation profession (including interpreting) for universities and institutions in Europe in the hope of converging and optimizing their TTPs. Among the competences is *Technical Competence.* It is specified explicitly (see below) in order to facilitate understanding and implementation.

Technical Competence (mastery of tools)
- Knowing how to use effectively and rapidly and to integrate a range of software to assist in correction, translation, terminology, layout, documentary research (for example text processing, spell and grammar check, the internet, TM, terminology database, voice recognition software)
- Knowing how to create and manage a database and files
- Knowing how to adapt to and familiarise oneself with new tools, particularly for the translation of multimedia and audiovisual material
- Knowing how to prepare and produce a translation in different formats and for different technical media
- Knowing the possibilities and limits of MT (p. 7)

The most inclusive and vigorous model of translation competence to date may be that of PACTE (1998) first proposed in 1998 and revised in 2003 followed by a series of experiments aimed to verify its validity (Lesznyak, 2007; PACTE, 2003, 2005, 2009, 2011). PACTE was founded in October 1997 by a group of translation teachers and translators from the Facultat de Traducció i d'Interpretació of the Universitat Autònoma de Barcelona. They set as their goal to investigate how translation competence is acquired "in order to create better teaching programmes, improve evaluation methods and unify pedagogical criteria" (PACTE, 2003, p. 43). Translation here refers exclusively to written translation and covers both direct and inverse directions (i.e. in and out of the foreign language). What is worth mentioning here is that PACTE is the first batch of researchers who claimed openly not only to propose a theoretical model of translation competence but also to validate it empirically with an appropriate research design.

In their initial model, the ability relating to tool use is couched in *the instrumental/professional sub-competence* which was defined as " the knowledge and abilities associated with the practice of professional translation: knowledge and use of all kinds of documentation sources; knowledge and use of new technologies; knowledge of the work market and the profession (prices, types of briefs, etc.)" (PACTE, 2003, p. 49).

As stated above, aimed at an empirically validated model of translation competence and its acquisition, PACTE followed their first proposal with a series of experiments using six language pairs and a combination of observing instruments such as PROXY (a kind of user-monitoring software), questionnaires and guided TAPs. After the first two exploratory tests conducted in 2000, needs arose not only to modify the measuring instruments but also to revise the 1998 model of translation competence. One of the changes made to the model is that *Knowledge about Translation Profession* which was ascribed separately to *Extra-linguistic Sub-competence* and *Instrumental/Professional Sub-competence* was now found to be of greater importance in the translation process, to such a point that could constitute an independent sub-competence. Thus in the revised model *Instrumental Sub-competence* alone rises to be one of the five sub-competences that compose the holistic competence of translators, as shown in Figure 2. 2.

The *Instrumental sub-competence* is defined by the group as follows:

> Predominantly procedural knowledge related to the use of documentation resources and information, and communication technologies applied to translation (dictionaries of all kinds, encyclopedias, grammars, style

books, parallel texts, electronic corpora, search engines, etc.). (PACTE, 2003, pp. 43-66)

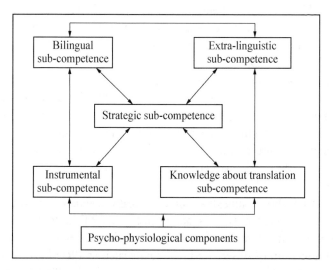

Figure 2. 2   The Revised Translation Competence Model by PACTE（2003）

Some other models, without a separate component of instrumental competence, alluded to or mentioned, if briefly, translators' *ability to use tools* among which the most often mentioned one is dictionary (e.g. Ágnes, 1983 and Campbell, 1991, 1998, cited from Márta, 2008, pp. 38 & 46; Gile, 2011; Reiss, 2000). Taking into account the time when these models are launched, that no reference is made to electronic tools and resources is not surprising.

As the literature shows, although the importance of the ability to use tools has been widely recognized by researchers, viewpoints still vary as to how it can be integrated into the model of translation competence, i.e. its place in the model and its relationship with the other components. Besides, another common problem facing all the above models is that they are all formulated top-down. That is, all the models are proposed based only on past experience or theoretical conjecture whose validity is subject to more empirical verification (Pym, 2003, p. 487).

2.2.2   *Empirical efforts*

Actually as early as decades ago, studies on translation process began to report differences in tool use between professional and student or novice translators. For example, Gerloff (1988, p. 106f.) found that professionals used reference works more frequently than students and bilinguals. While the students mainly consulted them to solve comprehension problems, professionals and bilinguals used them mainly for solving text production problems. Jääskeläinen's (1989) study revealed that novices looked up in dictionaries more frequently than advanced students and compared with novices advanced students consulted more dictionaries per problematic item. In addition, it was observed that novices preferred bilingual dictionaries, whereas advanced students favoured monolingual ones.

In recent years more experiments were designed to seek empirical data in support of the different models (e.g. PACTE, 2005, 2009, 2011; Qian, 2011; Wang, Mo & Liang, 2010). More evidence indeed began to appear that verifies the existence of instrumental competence and its positive correlation with the overall translation competence.

PACTE group prevails also in this aspect. In their 2000 exploratory tests, *consultation of printed and electronic materials* was already identified among expert translators' observable activities by way of direct observation and PROXY recording. Later in 2009 they continued to report what their following experiments had found about the three sub-competences they believed to exclusively belong to translation competence, namely, *strategic competence, instrumental competence* and *knowledge of translation.* This time the focus was on the dependent variable *decision-making.* The results showed that professional translators performed differently in the way they used tools from non-translators and translators' skillful application of translation tools in the process of translation could help them compensate for their linguistic shortcomings and attain better acceptability of their final products. Therefore, following this experiment, PACTE (2009) added a variable of *Use of Instrumental*

Resources which is believed to be related to the instrumental sub-competence, given that marked differences were observed between experts and non-experts.

Ronowicz *et al.* (2005, cited from Macken, 2010) found that experts engaged in less dictionary searches than laypersons. The finding was supported by Márta (2008) who probed deeper into the use of printed dictionaries in translation process. She did not only observed the number of dictionary searches but also the types of dictionaries searched and the depth of dictionary use by subjects with different degrees of translation experience.

Some research in translation teaching also lends indirect support for tool use as function of expertise of translation. Wang *et al.* (2010), with a constructivist point of view, carried out an experiment with college English learners in order to examine the effects of real-life translation tasks on the development of translators' instrumental sub-competence in college education. The results are confirmative with the findings showing that after training, students in the experiment group (1) used tools more frequently during the translation process, (2) solved problems more quickly, and (3) achieved higher acceptability in their translation products than those in the control group.

Different from all the top-down studies with a speculative theoretical framework as a starting point, Qian (2011) took a reverse direction and is, to date, the only one to have studied translation competence this way. Guided by the grounded theory, Qian tried to search, mainly via introspection, from real-life experience for the constituents of translation competence before theorizing it. Based on a series of open-ended interviews, semi-structured interviews and focus group interview, she filtered and organized the findings. Finally a pyramid model was established of translation competence constituents among which mastery of translation tools and IT resources was considered part of the operative skills.

2.2.3 A summary

All in all, empirical studies in light of translation competence are admittedly still few (Mu & Lan, 2011; Wang, Tang & Wang 2008). Those devoted to instrumental competence particularly are even less. However, the recent 10 years have witnessed an enhanced awareness of the need for more systematic and in-depth empirical research in this regard and an increased theoretical endeavor (as seen in the inclusion of tool use in those multicomponential models). Besides those studies reviewed that have been conducted in the realm of translation competence, many other studies, especially the increasing number of market surveys in view of the changed demands for qualified translators (e.g. Fulford & Granell-Zafra, 2005; Jia, 2013; Wang, C. Y., 2012) also shed light on the composition of translation competence.

Wang C. Y. (2012) is probably the latest market survey aimed to collect information about what professional qualifications of translators the workplace needs. 65 of the 120 surveyed enterprises of various sizes returned their responses. As Figure 2. 3 shows, 77. 3% of the respondents place importance on translators' command of TT and tools. This finding is consistent with many other surveys in terms of tool use. Only as time goes on, electronic tools and IT tend to overtake and even replace traditional forms of tools, such as printed dictionaries and reference materials on paper.

To sum up, data from theoretical research, empirical studies as well as social surveys concerning translation competence all point out to translator educators the fact that translators' ability to use tools, especially those computerized ones, has become increasingly important so that it cannot and should not be ignored in any training programme.

2. 3　CAT and translator education

Urged by the thriving role of TT in the marketplace, its relationship with translator education has become a subject of an increasing number of studies in recent decades. The following section will start with the arguments over the necessity for and significance of integrating TTs training into the translation curriculum at university level. Coming after them the status quo of the CAT

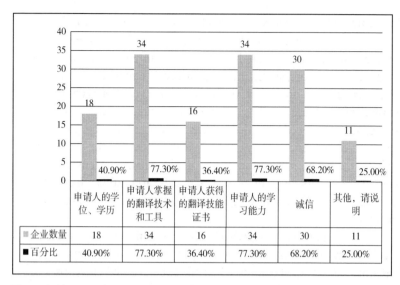

	申请人的学位、学历	申请人掌握的翻译技术和工具	申请人获得的翻译技能证书	申请人的学习能力	诚信	其他，请说明
企业数量	18	34	16	34	30	11
百分比	40.90%	77.30%	36.40%	77.30%	68.20%	25.00%

Figure 2. 3　Importance attached to the professional translator qualities by LSPs
(Wang, C. Y., 2012, p. 68)

training across the world will be depicted before a detailed description is made about some specific CAT courses offered in the academic context in search of the development trends and problems in CAT teaching.

2. 3. 1　CAT teaching: a brief history

As mentioned earlier, researchers (e.g. Haynes, 1998; Kenny, 1999; Kingscott, 1996 cited from Bowker, 2002, pp. 13-14; Schäler, 1998) have noticed the gap between translator training and the market demands as regards TTs as early as in late 1990s. They drew widely on technical, economic, pragmatic and academic evidence in support of their argument for bridging this gap. Warnings were long ago issued that if universities want their graduates to survive the competition of the twenty-first century, they have the obligation to equip their students with the skills to use CAT tools (Clark, 1994, p. 308) as data suggested that "graduates who are conversant with CAT technology are at a real advantage when it comes to working in highly technologised translation environments " (Kenny, 1999, p. 79).

As a result, in the following years, many universities started to offer CAT or CAT-related courses to trainee translators, such as Dublin City University in Ireland, Manchester University and Kent University of Surrey in the UK, the University of Joensuu in Finland and the University of Ottawa in Canada, Monterey Institute of International Studies in the US, and the Chinese University of Hong Kong, Beijing University, Beihang University and SYSU in China, to name just a few. One thing to note is that among these universities offering CAT training, the Chinese ones actually lag behind their American and European equivalents by nearly 20 years.

In China, although it has been over one and a half centuries since the first foreign languages school, *Jingshi Tongwenguan* (Peking Foreign Languages School), was set up in Beijing as early as in 1862, translation and interpreting studies as an independent discipline began only around three decades ago when the UN Translator Training Programme (TTP) was set up in 1979 in Beijing Foreign Studies University to provide professional training at post-graduate level (Wang, C. Y., 2012). But in response to the deepening of globalization and the accelerating social and economic development of China, TTPs mushroomed throughout the country in recent ten years, especially after the year of 2007 when the professional degree — Master of Translation and Interpreting (MTI) — was set up. The following table (Table 2. 2) quoted from Wang L. D. (ibid. , p. 56) gives us a clear picture of it.

In spite of the frenzy at the institutional level, the faculties on the teaching front are inevitably faced with a series of challenges in practice, on top of which is the design of the curriculum and syllabuses. Resolutions in this regard are urgently needed largely due to the recent reform in the translation discipline that differentiates between the professional degree and the academic one. The newly established BTI (Bachelor of Translation and Interpreting) and MTI are both professionally oriented and thus the traditional academically oriented curriculums and syllabuses are no longer applicable.

As the foregoing literature shows, one of the biggest changes in the translation profession is the computerization of its working environment. Since 2000, some forward-looking educators in the translation section of China have begun to call for more attention to the teaching of TTs in college-level translation programmes (e.g. Lv & Mu, 2007; Xu, 2006; Yuan, 2005). Their endeavor was rewarded with the managerial sanction of CAT as an elective in TTPs on both undergraduate and postgraduate levels in China (Zhong, 2007; Zhong, 2011). CAT as a course has since then been available in a growing number of universities in China (Wang, 2013). So today the problem is not so much about whether or not CAT should be integrated into the translation curriculum as what to teach about CAT and how to teach it.

Table 2.2　Establishment of TTPs in China（Adapted from Wang, L. D. 2012）

1979	Beijing Foreign Studies University
2004	Shanghai Foreign Studies University
2005	Guangdong Foreign Studies University
2006	3 undergraduate programmes (BTI)
2010	31 undergraduate programmes (BTI)
2007	15 postgraduate programmes (MTI)
2011	159 postgraduate programmes (MTI)
2013	106 undergraduate programmes (BTI)

Besides, benefits of incorporating CAT in TTPs are not limited to the improved curriculum and thus expectable better qualifications of prospective graduates. It is argued that the integration of CAT tools into translation curriculum can also create some new areas of basic and applied research such as translation pedagogy, human-machine interaction and the evaluation of technology (Bowker, 2002) as well as terminology, similarity, reusability, and translation quality (Chan, 2007, p. 5).

In a word, nowadays the inclusion of TT, be it in any form, into TTPs is no more an exception but a rule (Bowker & Marshman, 2009).

2. 3. 2   *CAT teaching: the status quo at university level*

Presently CAT training in college exists mainly in three forms: CAT as a degree/certificate programme at postgraduate level, CAT offered as a single course or a series of courses in translation/localisation programmes at both postgraduate and undergraduate levels, and CAT offered as professional courses for part-time social students.

A number of universities around the world have been found to be providing specialized programmes in CAT (though with the degrees/certificates phrased differently), such as Master of Arts in Computer-aided Translation in the Chinese University of Hong Kong, the Postgraduate Certificate in Translation and Technology, a vocational programme in London Metropolitan University, MA in Applied Translation Studies (MAATS) in University of Leeds, Postgraduate Certificate in Language Technology in Swansea University, MSc (Master of Science) in Translation Technology in Dublin City University in Ireland, the Translation & Localization Management program at the Monterey Institute in the USA and Master of Engineering in CAT in Beijing University in China.

More common, however, is that many universities with translation/localisation progammes have incorporated into their teaching plan one or more CAT-related courses depending on the staff expertise and school facilities. Of this kind, there are University of Leeds, University of Swansea, Imperial College of London, and University of Nottingham, Université de Genève, SOAS University of London, Kent State University, SYSU, Beihang University, to name just a few.

Although CAT training has been offered in many institutions in various forms, the training strategies and effects are still not satisfactory to those who need training (Wheatley, 2003 and Lagoudaki, 2006, cited from Bowker & Marshman, 2009, p. 64). Nevertheless, compared with other sources for CAT learning, such as translators' forum, vendors of individual tools and professional associations, institutional programmes enjoy at least two unparalleled advantages. For one thing they could provide more systematic and in-depth training to students who would stay in university four years in minimum; for the other, the training in university may give students a head start when competing in the job market, which may lead to greater successes on their later jobs. No wonder that many researchers call for better training of CAT technology in the academic context given the "sizeable gap to be filled" in here (Bowker & Marshman, 2009, p. 65).

2. 4　CAT teaching in higher education: a pedagogical gap

As Kiraly (2000, p. 124) noted, simply offering a course on technology does not ensure the desirable results, namely producing students with acquired technological competence good enough for the workplace. Pedagogy is therefore an important prerequisite to effective teaching of CAT but it has been largely neglected as the author found in previous research. The following section is aimed to reveal the undesirable gap in CAT pedagogy in both teaching practice and research.

2. 4. 1　A pedagogical gap in practice

In an attempt to know more about the pedagogy of the current CAT teaching, the author managed to collect some course descriptions via the Internet or personal contact. Among them six examples with enough desired details will be briefly presented before an analytical comparison is made to investigate their pedagogies.

2. 4. 1. 1　The teaching practice

1. Centre for Translation Studies (CenTraS) in UCL

The first example comes from Centre for Translation Studies (CenTraS) in UCL (University College London) which includes in the curriculum of MSc in Scientific, Technical and Medical Translation with a course (in their term *module*) named *Translation Technology*.

The course is described as below:

> In this module you apply your theoretical and
> conceptual background of translation to a very practical
> approach of using many current tools at the disposal of a
> translator. These range from terminology management and
> TM tools to advanced file formats and software
> localisation. The module includes hands-on experience with
> a wide range of translation tools in a Translation Lab with
> state-of-the-art facilities. Packages to be covered include
> the following: Wordfast Anywhere, SDL Trados Studio
> 2011^{TM}, OmegaT, memoQ^{TM}, and Alchemy Catalyst^{TM}.
> (Net 1)

Other than this, UCL also provides a series of standalone courses
in TT for social learners. Among the courses are *Project Management
for Translators, Introduction to MemoQ, Introduction to Subtitling,
Introduction to Dubbing*, etc.

For example, in the course *Project Management for Translators*,
emphasis in teaching is given to practical problem-solving skills in
real-world cases:

> You will then participate in a session designed to
> explore strategies and techniques which are useful for
> dealing with problems and issues which can and do arise as
> both a translation vendor and a project manager working in
> an agency. This will be a practical session using real-world
> examples. (ibid.)

In the course *Introduction to MemoQ*, students are taught about
not only functions of this software but also the "the rationale behind
the structure of memoQ in comparison to other software and the
reasons that have made it popular among translation companies"
(ibid.). A process-oriented view of translation is adopted and
translation project is incorporated into teaching as shown below:

> During the second part of the course, you will be
> introduced to the main functions of the software and you

will be guided through processes of translation, as well as building your own resources with memoQ. By the completion of the tasks that follow this process, you will be able to use memoQ for basic translation projects. (ibid.)

Yet the pity is that the exact way of teaching and assessing is not specified in the syllabuses and keeps us in wonder.

2. Department of Linguistics in SOAS University of London

The second example is from Department of Linguistics in SOAS University of London which offers a course also named *Translation Technology* to postgraduates. The syllabus of this course is the most complete.

This course is taught over 22 weeks with 3 hours' classroom contact per week through a whole year. The major topics covered in this course include (Net 2):

(1) Introduction to MT and online MT tools as an aid to translators
(2) Translation memory (TM) & TM management
(3) Terminology database (TD) & TD management
(4) Translation projects management
(5) Corpora (monolingual, parallel bilingual and comparable) as an aid to translators
(6) Principles and skills in localization (software and websites)
(7) Principles and skills in subtitle creation and translation
(8) Critical evaluation of MT and CAT tools
(9) Translation technology and freelance translators
(10) Critical review of TT and its trend

Software to be taught in this course ranges widely from SDL Trados Studio 2009, SDL Trados 2007, WinAlign, SDL Multiterm 2009, SDL Multiterm 2009 Convert, SDL Multiterm 2009 Extract, Déjà Vu X2, SDL Passolo, Sisulizer 2010, Alchemy Catalyst 9.0, to British National Corpus, AntConc and the subtitling set.

Students by the end of this course are expected to demonstrate the following learning outcomes (ibid.):

(1) familiarity with the major translation software available to them and be able to use the software to assist them in practical translation

(2) skills in terminology management and construction of small translation corpora or terminology bank to assist their translation practice

(3) the ability to make use of internet resources for their translation

(4) research skills in critiquing theories and principles on TT

A bit puzzling, however, is that students in such a practically oriented course is assessed by three essays only. Intentions behind it is unknown without further specification on the website.

> One essay of 3, 000 words to be submitted on the day of teaching, week 1, term 2 (35%); one essay of 2, 500 words to be submitted on the day of teaching, week 7, term 2 (25%); one essay of 4, 000 words to be submitted on the day of teaching, week 1, term 3 (40%). (ibid.)

3. Monterey Institute of International Studies

Another case is from Monterey Institute of International Studies where a technology component of courses are open to students beyond the MA in Translation & Localisation Management. It covers TM and MT, terminology management, software and web site localization, process standardization, IT/workflow strategies, and project management. Tools used in class include: Kilgray memoQ, SDL WorldServer, SDL Trados Studio, SDL MultiTerm, SDL Passolo, Google Translator Toolkit, Across Language Server, Lingotek Collaborative Translation Platform, and Systran. In these components there is a course of CAT that has become a compulsory one for all the students on translation and interpretation programmes.

As the CAT course objective states, this course is aimed " to provide students with a deeper understanding of state-of-the-art tools and processes in the field of Computer-Assisted Translation." (Syllabus of CAT 2, 2012). The teaching methodology includes "lectures, visual presentations, demonstrations, discussion groups, and immediate use of skills learned. Students will also have opportunities for individual/group presentations" (ibid.). What is worth mentioning here is that the syllabus stresses that "raising questions and points for discussion in class is strongly encouraged" (ibid.). The traditional form of assessment — quizzes and examinations — is adopted. Yet an additional component of students' portfolios is somewhat different than usual, which according to the syllabus consists of "items created as part of class activities" (ibid.).

4. Dublin University

The fourth example is from Dublin University which opens a course of *Translation Technology* to students from MSc in Translation Technology, GDip in Applied Lang & Intercul Studies and Master of Arts in Translation Studies. According to the syllabus found online (Net 3), the course is aimed to develop students' knowledge, skills and competence in the field of TT. Students in this module will learn specialist technologies used in the translation industry by attending lectures and having hands-on experience in lab sessions. Critical reflections on the values of the technologies are also encouraged.

5. School of Cultures, Languages and Area Studies in Nottingham University

The fifth one is School of Cultures, Languages and Area Studies in Nottingham University which offers a course named Technology Tools for Translators in two phases, with the first phase being a compulsory module for the MA in Chinese/English Translation and interpreting and the second phase only an optional one. The content of the first phase is as follows:

This module will enable students to obtain practical skills and comprehensive knowledge of computer assisted translation (CAT), machine translation (MT), the industrial procedures of translation and localization projects as well as recent developments. Professional and ethical awareness are to be raised to develop students' capabilities to liaise with clients and to work as in-house translators with considerable level of technological and project management skills. This module maximises practice by simulating projects of the real-life language service providers (LSP) to make full use of Trados series software to produce written translation in different digital formats (Doc, Xls and Ppt), website localization (Html), and some software localization (Exe). (Net 4)

The second phase of this module is aimed to "enable students to acquire key skills and knowledge to work as freelance translators" (Net 5). Therefore, this second phase focuses on the use of free translation tools designed for and developed by freelancers, including two computer assisted translation (CAT) tools — OmegaT and Across, and a project management tool — Project-open. Students will be made aware of and understand the nature, requirements and resources of freelance translation. So far Nottingham University seems to be the only one to have developed a course devoted to training freelance translators.

What is impressive in this syllabuses is that students' learning outcomes are categorized into four kinds, namely knowledge and understanding, intellectual skills, professional practical skills and transferable skills (Net 4).

The course is conducted through 11 weeks with workshop once a week, each lasting for 2 hours. The assessment method is a 2000 word coursework based on LSP simulation assignments.

6. Department of Languages, Translation and Communication in Swansea University

Lastly in Swansea University, there are two courses regarding TT. One is Translation Tools, a compulsory course for both MATLT/MATLTE and MATI/MATIE. The other one is Translation Technologies, an optional module for all the translation and interpreting programmes.

Translation Tools is a 20-hour guided class course. This course covers a wide range of applications, ranging from the translation-specific aspects of generic software (e.g. word-processing, spreadsheet), through online information retrieval (e.g. Web-based terminology mining, domain knowledge acquisition), to specialised CAT tools (e.g. Translation memory, terminology management functionality) and even MT evaluation. Two CAT systems are mentioned explicitly which are Trados and LionBridge. Students at the end of this course are expected to have "a detailed understanding and practical experience of some of the industry — standard Computer-assisted Translation (CAT) technologies increasingly in demand in the translation professions today". Assessment will be entirely based on students' performance in and reflections on a simulate project assignment which is introduced to the teaching process midway in the course. More specifically, "20% of the marks for the assessment are awarded for the quality of the data files submitted by the whole group, 20% for the quality of the translation and terminology work done by each language team, and 60% for an individual reflective report on the project" (Net 6).

Translation Technologies is targeted at students who are "interested in a career as in-house IT specialist and/or localizer, a developer and tester with one of the major translation tools companies" (ibid.) and covers, therefore, more tools (e.g. TM systems Déjà Vu X2 and StarTransit/Termstar NXT, software localization package Passolo 2011 and desktop MT system Systran 7) and entails more theoretical and reflective content (e.g. investigation, evaluation and comparison between technologies with different design philosophies, strengths and weaknesses). Students may choose as their mode of assessment between two kinds of comparative report in which they have to report on their testing routine in comparing different packages and the comparing results.

2. 4. 1. 2 Analysis and comparison

With different target students, the specific design of each course varies from one another to some extent. The teaching content especially is presumed to be subject greatly to local constraints, such as the financial condition, the staff expertise and the availability of CAT technologies. The observation attested our conjecture yet revealed nonetheless some commonality among the cases. To begin with, TM systems, MT and terminology tools and management are unanimously included. In addition, although CAT packages taught in the courses vary in brand and number, there is no exception to the inclusion of Trados, the leading brand of commercial CAT systems. More peripheral content involves internet retrieving skills, the use and even creation of corpora, and localization tools. This finding indicates to us that regardless of the constraining factors varying in different situations, designers of CAT teaching have reached an agreement at large on what to teach in such a course, which makes research into CAT pedagogy all the more an urgent need.

A probe into the selected examples indeed discloses to us some prominent features of CAT teaching concerning its pedagogical design. The analysis will be reported below in terms of the teaching objectives, the teaching material/instruments, and the teaching as well as the assessment methods respectively.

Firstly we can easily perceive the tacit agreement among the designers on their expectations of the course effects (phrased as "course objectives" or "learning outcomes" that revealed different teaching paradigms). The designers of the courses seem to share the common notion that CAT teaching in higher education should not be confined to the operational skills of a limited number of specific translation tools or resources. Words like " understanding ", "evaluation" and " reflections" are frequently used in the course description showing that equal attention is also given to the " philosophies" and " principles" behind the tools and students' ability to evaluate and judge critically the value of different tools or technologies. Among the examples Nottingham University stands out easily for its specific and well-thought categorization of the learning outcomes which, comparatively speaking, enjoys two benefits: one is for students taking this course who would feel well-informed of what they will get from this course; the other is that such a framework provides a clear basis for understanding the role of the CAT course in the curriculum of the programme.

Secondly, real-life translation cases and translation projects are adopted as the teaching instrument almost in every one of the above cases. It is apparent that a process-oriented view of translation has been unanimously held by the trainers as shown in the cases that to inform students of the real-life industrial process or workflow is highlighted in most of the courses.

In addition, the author also observes some disagreement in the selected cases, which suggests to us the problems awaiting solutions. The first problem concerns the teaching methods. A mixture of teaching methods are found among which there are workshop, demonstrations, lectures, hands-on lab classes, presentations and group discussions. Some universities like Nottingham University adopts only workshop for classroom teaching while some others adopt a blended method with the rest not even mentioning the method at all. The second one is about the assessment methods. Some courses employ simulate projects as the form of assessment, while some others opt for essays which seems a bit curious for such a practice-oriented course. Only one of the courses, namely that of Monterery Institute of International Studies (MIIS), has added a process dimension in assessment by including the student's portfolios in the final grading and others seem to be entirely reliant on the final exam, be it a project, a report or an essay. It has been a rule that the assessment should be aligned with the teaching objectives for the achievement of the student to be measured reliably. But among these cases, the author has strong reasons to doubt the validity and reliability of the choice of an essay or a report as the assessment instrument and the entire reliance of the assessment on the single final exam.

Besides, some clues can be found in the word choice in the above cases to help infer what pedagogy is adopted. In majority of the cases the designers use "course objectives" and "course trainer" or "instructor" which all fall into the traditional teaching paradigm with teachers as experts in class to transmit knowledge to students. In contrast, Nottingham University uses "learner outcomes" and "convener" and Swansea University uses "coordinator" which show a sign of student-centred teaching approach.

2. 4. 1. 3   A summary

To sum up, an analysis of these cases leads us to the following findings about their pedagogies:

(1) CAT teachers agree to a great extent on the desired learning outcomes of this course;

(2) CAT teaching designers seem to neglect largely the importance of pedagogy as revealed from these course descriptions given either the absence of teaching method specification or the vagueness in terms describing the teaching approach;

(3) The inconsistence between the teaching objectives and the assessment methods in most of the cases suggests to us also the course designer's ignorance of the importance of pedagogy or even worse lack of pedagogical knowledge;

(4) Although the teaching objectives of the sample courses have incorporated the new demands for modern translators, most of them may still take the traditional teaching paradigm which has been largely believed to "fail to produce translators who are capable of the flexibility, teamwork and problem-solving that are essential for success in the contemporary language industry, not to mention the creativity and independent thinking that have always been the hallmark of the finest translators" (Baer & Koby, 2003, pp. vii-viii).

2.4.2 A pedagogical gap in research

The recent decade has witnessed steadily growing attention to CAT teaching, a young subject in translation programmes, from researchers, especially teacher researchers in universities. It is yet far from enough. The author searched on CNKI for studies in China with *themes* of *translation teaching* (using key words "translation + teaching"), *CAT* (using key word "CAT") and *CAT teaching* (using key words "CAT + teaching") respectively and produced the results as shown in Figure 2.4.

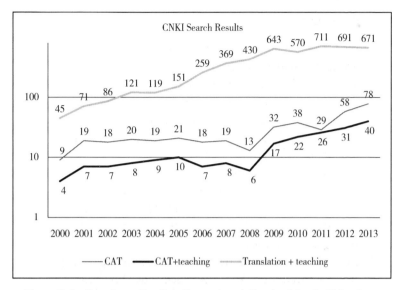

Figure 2. 4　Search results of studies on translation teaching in China from CNKI（2014）

In the chart we can see that studies of these three topics all increased enormously over the recent decade with an abrupt surge in 2009. The number of studies from 2000 to 2013 increases around 10 times. Yet comparatively speaking, being about one-tenth the total of the studies on translation teaching in general now, the number of CAT teaching studies is still too small. A deeper analysis of the statistics shows us that although the number of the published papers on CAT teaching grows by ten times, the proportion of it in the total of studies on translation teaching even decreases from 8. 9% in 2000 to 6% in 2013. Against the background that there are currently over 120 universities in the world which offer CAT-related course(s) (Wang, H. S., 2012) with teaching quality yet unsatisfied (Wheatley, 2003; Lagoudaki, 2006, cited from Bowker & Marshman, 2009), this finding about the present scarcity of published papers on CAT teaching, on the one hand, constitutes an undesirable loophole in the research and, on the other, makes it very hard for teaching practitioners to share and benefit from their collective experience in CAT training.

A broader look at the publications on CAT teaching worldwide brings us a more optimistic view with researchers in the US, Canada, and some European countries (prominently Spain) contributing a variety of papers and monographs to this field. All in all, direct contribution to CAT teaching comes mainly from five sources: one is the textbooks with translator trainees and trainers as target readers who need to tackle the tools training (e.g. Austermühl, 2001; Bowker, 2002; Qian, 2011; Quah, 2008; Xu, 2010b). Another one is the academic papers devoted wholly or partly to CAT teaching design (e.g. Luo, 2010; Qian, 2009; Samson, 2010; Sánchez, 2006; Xu, 2010a; Zhang, 2010). Also illuminating for the CAT teaching are the published market surveys aimed to explore the uptake of TTs in the profession (e.g. Fulford, 2002; Fulford & Granell-Zafra, 2005; He, 2010; Hou & Yan, 2008) as well as systematic discussions about definition and classification of TT (e.g. Alcina, 2008; Hutchins & Somers, 1992; Melby 1998; Vilarnau, 2001, cited from Alcina, 2008). The last but not the least is the discussions about CAT pedagogy (see Alcina, Soler & Joaquín, 2007; Colina, 2009; Cui, 2012; Davies, 2010; Luo, 2010; Samson, 2010; Pym, 2012; Sánchez-Gijón *et al.*, 2009; Xu, 2010a; Zhou, 2013).

However, of the above-mentioned five classes of studies, the last one, namely that on CAT pedagogy, is still smaller than the other four classes and lacks systematicity and depth in content. This conclusion is made based on the finding that among all the studies covering CAT pedagogy, none is found to be wholly devoted to this issue. Moreover, most of the researchers base their arguing for or proposals of teaching methods only on their personal experience, without or without reporting any methodological justification or theoretical basis in education, psychology or cognition.

2. 4. 3 A summary

As revealed in Sections 2. 4. 1 and 2. 4. 2, how to teach CAT in higher education was and is still an issue largely neglected in not only teaching practice but also academic research.

This pedagogical gap in CAT teaching is, for one thing, due to the overall neglect of the theoretical importance of pedagogy for translation teaching practice in the realm of translation studies (Kiraly, 1995; Cronin, 2010) and for the other, due to the short history of CAT being an independent subject/discipline in higher education. Whatever the causes, the lack of pedagogical research will presumably hinder a healthy development of the CAT teaching preventing CAT teachers from benefiting from their mutual experience on the whole and impair the quality of the teaching outcomes in terms of not only the CAT skills but also the general translation competence. Just as Baer & Koby (2003) pointed out, for translator education, "the *how* is as important, if not more so, than the *what*" (p. viii). Maria-Luisa Arias-Moreno (1999, cited from Baer & Koby, 2003) claimed even more directly that "if the translator has no formal training [in translation pedagogy], the experience is more than chaotic and catastrophic for students" (p. viii).

2.5 A promising innovative approach to CAT teaching: PBL

2.5.1 *The constructivist turn in translation pedagogy*

Recent years has witnessed a constructivist turn in pedagogy worldwide from teacher-centred knowledge transmission to student-centred learning construction under the influence of writings of such eminent philosophers and educators as Lev Vygotsky, John Dewey, Jean Piaget (cognitive constructivism), Jerome Bruner (social constructivism) and Richard Rorty among others. Alongside this paradigm shift, the traditional teaching approach with the teacher as the sole authority in the classroom imparting knowledge to students who can but passively receive it is now giving way gradually to a more interactive model where students play a much more active role and thus take more responsibilities in constructing their knowledge.

This revolutionary approach to pedagogy was inevitably ushered into translator education and has been gaining momentum in its development as it fits well with the modern demands from both the educators and the professional world of translation. In fact criticism against the traditional classroom-bound teacher-centred instruction in translator education has also long before been issued. The criticism mainly stems from the inadequacy of traditional pedagogy for meeting the two-fold end goals of translator education, namely, to satisfy the needs of the job market on the one hand and, as part of higher education, to produce motivated independent life-long learners on the other. For example, in discussing the educational goals of translation programmes, Gabrian (1986) reminded us: "Universities are unlike schools in that it is not their task to spoon-feed knowledge to students to be memorized (but not digested) and regurgitated for exams. Rather, the main task of universities is to encourage students to think and act responsibly and independently" (p. 54). He pointed out that the foundations of professional translation are precisely responsibility and independence (p. 54). Olvera-Lobo *et al.* (2007), against the modern working conditions of professional translators, pointed out that the traditional teacher-centred teaching is obsolete and student-centred alternative is more favourable which fosters student autonomy and eliminates " the figure of all-knowing teacher capable of resolving all problems " (p. 520).

Kiraly (2000) has been widely cited as a milestone in didactics of translation in that it is the first systematic endeavour to inform translation skills teaching with research findings from translation studies, foreign language teaching as well as psychological and cognitive studies "for bringing innovation to curriculum and syllabus design, and to pedagogical procedures in translator education" (Kiraly, 2001, p. 50). Kiraly (2000) noted in the book that while the transmissionist view of translation (that conceptualizes "translation as a process of transferring meaning from one text to another") has given way to "the perspective that readers (and translators) make meaning as they interact with texts", the traditional transmissionist approaches to the training of translators remain predominant in current translation classrooms. Meanwhile he argued that different from the past, " translators today cannot afford to be linguistic hermits, sitting alone behind a typewriter and surrounded only by dusty tomes. Translators are embedded in a complex network of social and professional activity" (p. 12). Such observation of the gap between research findings and teaching practice, and of the teaching programmes' failure to take into consideration the changed professional profiles of translators begot the author's reflections and investigation, with the book as the result.

A major contribution of this work is, as the title suggests, the introduction of social constructivism into translation teaching methodology which has had a far-reaching influence on translator training largely. Kiraly (2000) argued that the social constructivist approach is particularly suited to translator education since translator competence can be seen as "a creative, largely intuitive, socially-constructed, and multi-faceted complex of skills and abilities" (p. 49). Among others, the underpinning notion Kiraly brought to translator educators is that learners should not be expected to reach an objective truth via education; rather they can but "create or construct meanings and knowledge through participation in the interpersonal, inter-subjective interaction" (p. 50). Along this line, the current pedagogical goals, learning environment, classroom management, teacher-student roles and assessment methods will all have to be reformed to not only comply with the cognitive principles of learning but also cater to the social needs. According to the qualitative research project Kiraly (2001) carried out to investigate the role the constructivist epistemology has in the design and implementation of translation instruction, "a constructivist approach does indeed lend itself to the education of translators" (p. 53). He found that "the approach inherently promotes a more equitable distribution of authority in the classroom and higher levels of motivation and active participation. It also encourages students to assume responsibility for their own learning." (ibid.)

2.5.2 *The best exemplar of constructivist pedagogy*: *PBL*

The relevance of constructivism with learning and teaching in higher education has since long been emphasized and explored (e.g., von Glasersfeld, 1995, 1998; Brooks, 2002). One of the most popular

48

teaching approaches derived thereof is the problem-based learning (PBL) theory which has been widely welcomed and verified in professionally oriented programmes worldwide. PBL is a learner-centred teaching model initiated by Canada's McMaster University Medical School in the 1960s and has since then adopted and further developed in medical schools throughout North America. This approach has been mainly used in medicine and business schools and, given its success in promoting professional competence, was later introduced into a wider range of fields including science education (Barak & Dori, 2005), engineering education (Rohani, Hassan, A., Hassan, S., Hamid & Yusof, 2005), educational administration (Macrae, 2000) and in recent years also to language teaching (Larsson, 2001) and translator education (Huang & Wang, 2012a, 2012b; Inoue, 2005; Kerkkä, 2009; Sánchez-Gijón *et al.*, 2009; Stewart *et al.*, 2010). A fuller review of the history of PBL application can be seen in Hung *et al.* (2008). PBL is considered by many researchers to be the most innovative instructional method to date (Hung *et al.*, 2008, p. 492).

With the widespread application world-wide across different disciplines and at different levels of education, PBL has been found to be used on at least three levels, namely, as theoretical learning principles, educational models and education practices. Nevertheless, it is almost unarguably accepted that PBL is characterized by using ill-structured authentic problems as stimulus of the learning process which is self-directed with tutors acting as facilitators and students learning in collaborative group work (e.g. Barrows, 1996; Graaff & Kolmos, 2003; Savery & Duffy, 1995; Taylor & Maflin, 2008).

Among translation studies, Inoue (2005) seems to be the most comprehensive discussion so far about the feasibility of PBL to translator education in general based on both theoretical argument and empirical exploration. His investigation of the research on the differences between experts and invoices led to the finding that it is the manner of problem-solving and the self-regulation or reflection involved herein that distinguishes expert from novice translators. Following it the author assumed that " translator education would benefit from the incorporations of problem-solving in order to bridge gaps between a novice and an expert" and thus proposed that "it is sensible to employ a pedagogical approach for novice translators in which solving problems is the central focus" (p. 8), namely PBL. Moreover, the pilot study Inoue carried out later was received very favourably by translator trainees which, regardless of the constraints of its scale and research methodology, offers positive evidence for PBL's applicability in the context of translation teaching.

There have also been attempts to integrate PBL into courses with specialization in translation programmes. Sánchez-Gijón et al. (2009) reported "how terminology courses may be adapted to current European higher education requirements using PBL methodology" (p. 105). They argued for the suitability of PBL for the terminology course on two accounts: on the one hand, PBL is a learner-centred teaching approach which would be particularly suited to meeting current European higher education requirements for students to be more active participants in the learning process; on the other hand, PBL, as a method of instruction that encourages students to "learn-how" by working collaboratively in groups to deal with real-world problems, can develop students' problem-solving skills that are necessary to meet the challenges posed by terminology work in their day-to-day activities as professional translators.

Huang and Wang (2012a, 2012b) investigated the applicability of the PBL approach to English translation and interpretation classes respectively in a university in Taiwan. In both studies, the students' perspective was taken in measuring the effectiveness of integration of the PBL methodology. In the translation class (Huang & Wang 2012a), students' reports on the four attributes, namely satisfaction, suitability, motivation, and self-achievement, were explored after the students had been exposed to a semester of courses administered in the PBL approach. In the interpretation class (Huang & Wang 2012b), students' attitudes, satisfaction, motivation, and self-achievement were investigated qualitatively in examination of the feasibility of the PBL methodology. Both studies reported positive evidence for the effectiveness of PBL as it "has provided a rewarding and quality learning experience for students" (p. 14). The students' motivation and self-achievement have been reported to be greatly enhanced although students were not unanimously satisfied with the approach. In light of the mixed attitudes towards PBL in translation class, the researchers suggested offering more peer-evaluation, on-line meeting systems, and supervision over students' progress in implementation to improve the effects.

Stewart *et al.* (2010) introduced the application of PBL in the translation project Teaching Medical Translation (TMT) at University of Heidelberg in exploration of its effects on translator trainees' learning process and acquisition of translation knowledge and skills. Their project adopted a problem-based cooperative approach to translation that was designed to revolve around "authentic and complex problem scenarios, realistic working context, a self-directed learning process among students working in small groups, and authentic learning resources and reference tools" (p. 8). The results of this project seemed to be very promising as evidenced by the highly positive impacts of PBL on students' affection (e.g. confidence, motivation, group solidarity), interpersonal and communicative skills (e.g. the ability to reach compromises, to cooperate, to tolerate different working styles) and the quality of the translation product (see Stewart *et al.*, 2010, p. 25 for a full list of comparative advantages of PBL over tranditional teaching methods). Revelant to the current study is his finding that by this cooperative method "students acquire confidence in the use of electronic and other media, as well as with computer programs and tools designed specifically for translators" (p. 23).

Although the limited number of attempts to integrate PBL into translator education does not allow us to be assured of the overwhelming advantages of PBL over traditional teaching methods, the prevalent application of PBL in many practice-oriented training programmes all over the world and the illuminating effects reported in the previous research in translation teaching make it at least very inviting to know to what extent the known positive effects of PBL in, for example, a medicine classroom will also be visible in a translation classroom. In addition, Savery and Duffy (2001) thought "PBL one of the best exemplars of an instructional model with a clear link between the theoretical principles of constructivism, the practice of instructional design, and the practice of teaching" (p. 31), which reassures the author about its readiness for application.

2.6　A summary

In this chapter, the author has clarified the CAT-related concept, explored the theoretical basis of CAT skills in translation competence studies, and described briefly the history and the status quo of CAT teaching in higher education followed by a closer look at the available research in this regard. This part is intended to justify the necessity of more attention to CAT teaching in academic context and warrant more specifically a research-informed and empirically based study to address the "pedagogical gap" in the current research on CAT teaching. The constructivist turn in pedagogy which has been proven to be a promising direction for higher education provides a valuable lead for the current pedagogical exploration. Further, PBL as the best exemplar in this paradigm shift has been introduced by many researchers into translator education at tertiary level and has shown highly encouraging and positive effects on the learning outcomes. This finding, combined with the envisioned advantages of its adoption to CAT teaching, makes it a justifiable goal to testify PBL's applicability to the CAT course in higher education.

Chapter Three
The Research Design

After the attempt to explore a PBL approach to CAT teaching is preliminarily justified, this chapter will be devoted to an elaboration on the research design. The methodology will be clarified below in light of the research aims. Specifically, the research questions will be formulated before the research type is decided and the researcher's theoretical affiliation to constructivist and interpretive research stated. Following it the research methods will be chosen for each research procedure with research strategies and instruments introduced respectively in detail.

3.1　Research questions

Stimulated by the findings in Chapter Two, the author intended to pioneer an exploration for a pedagogical innovation in CAT teaching with a PBL approach to fill in the void that has been revealed therein. More specifically, drawing on the recent pedagogical development towards constructivism and inspired by the discussions about feasibility of PBL in translation didactics, the author, with her 6-year experience in teaching CAT, has envisioned the extraordinary power of PBL to enhance the quality of CAT instruction and therefore intended to introduce PBL into CAT teaching in TTPs as a promising alternative pedagogy.

As the first attempt to introduce this approach into CAT teaching practice, the author would have to build the connections between PBL and CAT teaching, in search of both theoretical and empirical evidence for its applicability. Therefore, the author initiated a project of educational design research at SYSU with a hope to answer the following questions:

(1) How is PBL aligned with CAT teaching at the tertiary level?

(2) How can PBL be applied to the design of the teaching approach to the CAT course?

(3) How does the PBL approach work in its preliminary implementation in the CAT course at SYSU?

In a word, the aim of this study is to build connections between PBL and CAT instruction both in theory and in practice, establishing the basis for more systematic design and inform further implementation of CAT instruction with the PBL approach in higher education.

3.2 Research type

Akker *et al.* (2006) and Institute of Educational Sciences *et al.* (2013) are used as reference to decide the research type of the current study. In light of the research questions formulated in the previous section, this study can be categorized as educational design research across exploratory and design & development stages.

Akker *et al.* (2006) attempted to demarcate and define the emerging trend of educational design research which has shown its value for researchers and educators in bridging the persisting undesirable gap between educational theories and practices but is still characterized by "a lack of consensus on definitions". Among all the arguments, Barab & Squire's definition (2004) is found to be a generic one that encompasses most variations of educational design research: "a series of approaches, with the intent of producing new theories, artifacts, and practices that account for and potentially impact learning and teaching in naturalistic settings" (p. 2). To further clarify this type of research, five defining characteristics were formulated drawing widely on previous studies, which are as follows:

(1) Interventionist: the research aims at designing an intervention in the real world.

(2) Iterative: the research incorporates a cyclic approach of design, evaluation and revision.

(3) Process-oriented: a black box model of input-output measurement is avoided; the focus is on understanding and improving interventions.

(4) Utility-oriented: the merit of a design is measured, in part, by its practicality for users in real contexts.

(5) Theory-oriented: the design is (at least partly) based upon theoretical propositions; and field testing of the design contributes to theory building.

This study falls neatly into this type of research as it is intended to design a new pedagogical intervention in CAT teaching in higher education, in hope of not only enhancing the teaching quality but also contributing in turn to a better understanding of the theory of PBL.

Institute of Educational Sciences *et al.* (2013) claimed to create "a common vocabulary to describe the critical features of ... study types to improve communication within and across the agencies and in the broader education research community" (p. 4). It divides research vertically up into six stages, ranging from early knowledge-generating projects to studies of full-scale implementation of programs, policies, or practices which provided a broad framework that clarifies the purpose, justification, design features, and expected outcomes for each type.

Specifically, among the six stages of research in this report are foundational research, early-stage or exploratory research, design and development research, efficacy research, effectiveness research, and scale-up research, following roughly " the logical sequence of development of basic knowledge, design, and testing" (ibid., p. 4). Yet as noted later in the report, understanding and knowledge do not necessarily develop in a linear way. Thus actual studies may incorporate elements from different types or skip across the stages. According to the report, *exploratory research* is " to investigate approaches to education problems to establish the basis for design and development of new interventions or strategies", while *design and development research* " draws on existing theory and evidence to design and iteratively develop interventions or strategies, including testing individual components to provide feedback in the development process" (p. 9). These two stages of research combined fit quite well with this research's purpose of introducing a new instructional methodology into CAT teaching as informed by innovative learning theories.

To sum up, the current study, in accordance with its research questions, falls neatly into educational design research spanning the two stages of exploratory research and design & development research.

3.3 Research methodology

3.3.1 Research paradigm

As Schensul (2012) pointed out, scientists conduct their research under influence of different understandings of the objects of their study. For example, among the four influential paradigms in the social sciences he introduces, namely positivist, interpretivist, critical, and participatory, positivists believe that "reality is external to the self, that it can be observed, and that the tools used in the conduct of research can produce information that is reproducible and potentially replicable if collected under similar circumstances" (p. 76), while researchers taking *interpretivist* position believe that social and cultural phenomena present themselves by way of meaning constructed by the actors experiencing them, and along this line interpretivists would choose to interact with the participants to understand the phenomena from how they interpret them. It is, therefore, necessary to make clear the researcher's choice of paradigm before specific methods could be decided. Meanwhile, some research combines interpretivist and positivist approaches, falling into the mixed-paradigm research, "highlighting the voices and views of the participants, in interaction with the results and interpretations of the researchers" (p. 77).

The current study was set in the interpretivist paradigm as determined by its being an exploratory study on the one hand, and the constructivist philosophy underpinning PBL on the other. That is, knowledge is constructed by individuals in interaction with others and his/her social context by way of interpreting and evaluating. The consideration was that firstly the guiding research paradigm should not run counter to the conception of knowledge underlying the theory of PBL. Besides, at the very initial stage of introducing PBL into CAT teaching practice, there were not pre-identified variables of which a relationship of causality or otherwise to be discovered quantitatively. The major goal was to understand how PBL worked as perceived by the actors in the implementation process which would in turn inform later adaption and formal application. So for the time being, the best the researcher could do was stay open to what emerged in the process of implementation and construct her understanding accordingly.

The situation described above makes, therefore, interpretivist paradigm the perfect choice for the current study. Subsequently, adoption of such a paradigm has at least the following four implications for the test methods to be adopted:

First, qualitative research would be more preferable to quantitative one as the meanings each individual participant attributes to his/her experience in the course of the study may be better explored and understood.

Second, the research participants' experience and attitudes may be the best starting point for the researcher to understand the effects of introducing the PBL approach to CAT teaching.

Third, the researcher does not have to be an outsider staying away from the object under study. Instead, the researcher may as well be involved in the research process so as to fully understand it as there is no "reality" independent of the person who perceives it. The researcher is supposed to be part of the process and come to understand the function of PBL on teaching and learning in interaction with the study participants.

Last, the research findings may not be readily generalizable to wider contexts as what is normally expected of traditional scientific studies. Yet the current study can hopefully provide more details or new questions about the learning process with the PBL approach which could possibly deepen our understanding of the learning approach or furnish new directions of study on this subject.

3. 3. 2 Research procedures

R & D was referred to when the author decided on the exact steps to be taken. According to Gall *et al.* (2003, p. 569), R & D is an industry-based development model in which the findings of research are used to inform designing new products and procedures. Systematical field-testing and evaluation follow normally against a set of criteria of effectiveness which will provide suggestions to refine the design until they meet the standards. R & D is widely adopted and acclaimed by education researchers because it bridges the basic and applied research findings and teaching practice by translating them into usable educational products. It was therefore employed by the author in specifying the exact steps to follow in this research.

Borg and Gall (1989, pp. 784-785) outlined eight steps in the Research and Development Cycle. Given the fact that the current study is exploratory research making initial attempts at building the connection between PBL theory and CAT teaching practice, it suffices for the research objectives to accomplish the first five steps. They are:

(1) Research and information collecting — including a needs assessment and review of the literature;

(2) Planning — establishing a set of specific objectives for determining what the eventual product should achieve;

(3) Development of a preliminary form of the product — including formation of instructional materials, procedures and the evaluation instrument;

(4) Preliminary field testing — interview, observational, and questionnaire data are collected and analyzed;

(5) Main product revision — revision of product as suggested by the preliminary field test results.

For convenience of illustrating the methods adopted, the author grouped the five steps into three stages. That is, steps (1) and (2) compose the first stage, named Building Theoretical Connections, step (3) makes the second one, named Pilot Design, and steps (4) and (5) make the third stage, named Preliminary Field Test. Methods of the three stages will be specified below one by one.

3. 3. 3   Research methods

In this section, methods for each stage will be elucidated along with which the three general research questions are further specified.

3. 3. 3. 1   Stage one: building theoretical connections

At this stage, the author relied heavily on literature review to build the connections between PBL with CAT teaching. Robinson & Reed (1998) defined a literature review as " a systematic search of published work to find out what is already known about the intended research topic" (p. 58). Furthermore, Leedy (1989) pointed out that more knowledge of the studied problem will deepen researcher's understanding of it. Among the 23 benefits Onwuegbuzie, Leech & Collins (2012) identified of conducting a quality review is "identify relationships between theory/concepts and practice " (p. 1). Therefore, the purpose of a literature review was adopted not only to identify and analyse information relevant to the theories and practice of PBL, but also to help the author gain insight and understanding into the problem at hand so as to inform the bridging between the PBL theory and the CAT teaching practice.

EPPI-Centre (2010) suggested that reviewers use review questions as guidelines as to what to be undertaken in the review. Given the purpose of the review to answer the first research question, namely to align the PBL teaching methodology with CAT teaching at the tertiary level in China, the author broke it down into two guiding questions for the review:

(1) What is PBL?
(2) In what ways is PBL aligned with CAT teaching in college-level TTPs?

In order to systematically review the available papers relating to this research, a set of parameters of literature were identified respectively by referring to the methods proposed by the EPPI-

Centre (2010).

The parameters for the literature of the first focus are as follows:

(1) Written in English and Chinese;

(2) Worldwide origins;

(3) Focused on PBL theory with its application in instructional design and outcome evaluation;

(4) Databases searched: IEEE/IET Electronic Library, JSTOR, John Wiley, CALIS, Springer-Link Journals, ProQuest Research Library, Taylor & Francis Online, Elsevier, CNKI, Google Search (for open access resources);

(5) Dated 1995 onwards (with exception only to those classical papers by PBL founders).

The literature was searched for with the following key words:

PBL
+ pedagogy
+ curriculum/course /module/instructional design
+ implementation
+ effects/perception
+ evaluation
+ translation
(Key words being PBL alone or PBL plus any term on the right column)

Worth mentioning here is that seeing that PBL-related literature is widely distributed across disciplines, the author did not confine her search within certain periodicals or journals. Instead the well-known electronic databases accessible to the author via the university library (as specified above) were fully used. As pointed out by EPPI-Centre (2010), searching of databases may not locate all research reports, due to the limited inclusion of any databases. It is important therefore to use a combination of the available approaches, such as using general search engines like Google or hand searching some of key journals. So Google was then added to the searching tools and was found very helpful indeed for the author to get freely accessible materials especially many PBL application cases on university websites. More publications were further identified from the bibliographies of the papers acquired as long as they were found relevant to the current research.

Then, due to the immensely huge number of PBL-related publications, the author found it impossible to achieve exhaustiveness in gathering the relevant literature. In order to strike a balance between the inexhaustible number of publications and a comprehensive understanding of the subject, the author set up a step of screening, getting a core research base which consists of studies of meta-analysis, review and commentary. The author constructed her understanding of the development of PBL mainly in reliance on the core base, referring to the other available studies only for necessary backup information.

3. 3. 3. 2 Stage two: the pilot design

This stage of research is devoted to answering the following questions:

(1) What are the basic elements in the PBL instructional design model?
(2) How can it be applied to the design of a PBL approach to the CAT course at tertiary?

More specifically, at this stage, the author formulated a prototype PBL design framework which was later contextualized in a specific context, applied to a CAT course in undergraduate translator training programme.

This section will then report the methods used at this stage. Firstly, the establishment of a prototype design framework [i.e. Question (3)] was reliant on literature review too, which has been explicitly illustrated above in Section 3. 3. 3. 1 and will not be repeated here.

Then for its application to the CAT course in particular, constructive alignment (CA) (Biggs, 1996), as a course design approach, was adopted given its consistency with PBL's underlying constructivist philosophy. It emphasizes, too, the central position of students in the learning process and underscores that learners should be given ample opportunities to actively and cumulatively construct their own meanings (Biggs, 1996). Therefore, " representing a marriage of the two thrusts [constructivism and instructional design]" (ibid., p. 247), CA comes up as the best model for the current PBL course design.

The basic idea of CA is that the course has to be designed in such a way that the learning activities and assessments are aligned with the learning outcomes expected by the course, to guarantee the consistency in the teaching system. CA stands out easily from other models seeing it does not only furnish an "idea"; rather to bridge the "hiatus" most teachers have, it manages to operationalize the abstract "aims" and make possible workable teaching objectives, or, in his words, achieving descending "from the rhetoric of their aims to the specific *objectives* of a given course or unit" (ibid., p. 351).

More exactly, this stage of research will follow the three-step CA process which begins with clarifying the course learning objectives. Then in accordance with the objectives the assessment methods and teaching & learning activities are developed. The three steps form a dynamic cycle (Figure 3. 1) which situates the course in a specific context, draw together the learning processes (Knight, 2001) and sequence the learning content (Meyers & Nulty, 2009).

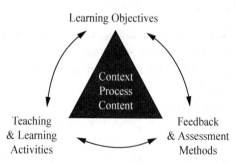

Figure 3. 1   The dynamic circle of CA process (Net 7)

In formulating the learning objectives for the course, the revised version of Bloom's taxonomy (Anderson & Krathwohl, 2001) was used as a benchmark for ranking the cognitive levels of the objectives.

In sequencing the teaching content, the model of zone of proximal development (ZPD) was adopted. As reviewed above, in PBL instruction, the learner is believed to bring with them different levels of prior knowledge. By accommodating newly acquired information or assimilating it into the existing knowledge, learners construct their new knowledge. Along this line, ZPD can guide tutors in their decisions about where to begin their instruction and what help to offer to facilitate students' learning. That is, the problems adopted should fall into the initial ZPD of most students and guidance should be provided as needed to guarantee their reaching a new level of ZPD. Moreover, with the learning process starting with a problem, PBL is an instructional strategy with a built-in requirement for students of diagnostic assessment of ZPD. Students in PBL instruction would have to assess what they have known about the problem and what more they need to find out, thereby identifying their learning issues which ideally direct, with or without tutor guidance, toward the anticipated learning outcomes by the instruction. Given individual differences in group learning, multiple ZPDs are advised to be contained in the classroom (Brown, 1994). That explains the adds-on benefit of PBL that it can accommodate differences among individual learners, allowing for individualized learning experience in authentic social interaction with tutor guidance and peer support. The teaching provided this way is termed *scaffolding* by Vygotsky (Campbell, 1999, p. 14). In other words, the teaching content is sequenced and scaffolded in such a way that facilitates learners' constructing their knowledge along their ZPDs.

3. 3. 3. 3　Stage three: the preliminary field test

3. 3. 3. 3. 1　The method choice

The purpose of this test was to understand how the pilot design works in real teaching in search of solutions to further improvements. It presupposes the following requirements for the method to be adopted. First of all, the epistemological notion underlying the method should be consistent with the understanding of reality inherent in this study pertaining to PBL. That is, reality is complicated and socially constructed rather than something that preexists just to be discovered. Secondly, since *educational design research* is characterized by the two prominent features of being context-bound and being focused on understanding rather than a simple judgment, this part of study was intended to seek more of the insiders' view in order to, rather than obtain the result changes, understand the process of these changes and how learners perceive such changes (Chen, 2008). To find more about how changes take place and how they are perceived by learners in this particular context is much more valuable for designers than to produce generalizable data in informing further improvement of the future design. This is especially true when quantitative research at this stage can hardly produce reliable and worthwhile data given the very limited time of the intervention (Akker *et al.*, 2006).

With all the above features taken into consideration, *qualitative case study* came up to be the most suitable method as McMillan and Dwyer (1989) had suggested. As Freebody (2003, p. 81) pointed out, "the uncertain, complex, messy and fleeting properties of educational contexts" render the traditional quantitative approach somehow ineffective in describing, understanding and explaining them. Case studies, however, with a strong sense of locality or a particular instance of educational experience, focus on the particular group of learners in a particular place and under particular conditions. Its great potential to unearth how the participants perceive the educational process makes it a perfect choice as the method of the initial stage of instructional design research (ibid., p. 81).

To sum up, at this stage, the author attempted to explore how the PBL approach works as perceived by the participants by a *qualitative case study*, or, more exactly, an *instrumental case study* (Stake, 2005). Sliverman (2006) also stressed the ability of case study research to open the way for discovery which may direct further inquiry to be made subsequently. Thus the findings of this study were meanwhile expected to shed light on possible further studies on the application of the pedagogical theory of PBL to CAT teaching.

3. 3. 3. 3. 2   Bounding the case

1. The research questions

As Moore, Lapan & Quartaroli (2012) noted, almost no educational programmes can be evaluated in its entirety. Basically an evaluation phase, this stage of research, adopting case study methodology, would have to "bound the case", that is, identifying the case chosen and clarifying what to be investigated in the case and what not before specific methods of data collection and data processing can be decided. In this section, therefore, the author will report the "bounding" of the case by clarifying the research questions, the time frame to be included, and other elements of the case related to the investigation.

First, this stage of research was intended to address the third research question which is broken down to the following three specific ones:

(1) How does the PBL approach work in the CAT course at SYSU as perceived by the participants?
(2) What are the pedagogical implications of the findings for its further application?
(3) What need to be further studied as revealed by the findings for an improved application of the PBL approach to the CAT course?

Being the initial effort to introduce PBL into CAT teaching, this stage of evaluative study was neither aimed nor likely to examine the relationship of causality among different previously identified variables. Therefore, as the first question shows, how the participants who have personally experienced this change of teaching methodology receive it was reasonably chosen to be the best starting point for the evaluation of this new approach. Besides, the in-depth exploration of the participants' reaction toward this approach during the case study was expected to reveal valuable information about its feasibility in CAT teaching, hence question (2), and point out the directions of further studies on the theory of PBL, hence question (3).

2. The case and the participants chosen

As the author's interest in PBL arose from the rising importance of and new social challenges for CAT training in TTPs at the tertiary level, the case chosen for the current study was thus defined by the course implemented with a PBL approach. As Stake (1998) states, what distinguishes case study from other research methodologies is its interest in the "case(s)" rather than "the methods of inquiry used". He also points out that the selection of the case should maximize the "opportunity to learn" (Stake, 1995, p. 451). Along this line, the author adopted purposeful sampling to a possible information-rich research site which met the following criteria and thus could provide the best answer to the research questions for the time being.

(1) First, the university the researcher would select should have a stable CAT infrastructure. In other words, the university had to have laboratories equipped with enough computers installed with main-stream CAT packages, such as SDL Trados.

(2) Second, the laboratory should be maintained regularly and effectively so that the computers and software are functional for effective CAT course delivery.

(3) Third, most importantly, the university had to have integrated CAT into the curriculum as an eligible course and be well-resourced to have sustained the offering of the course as planned.

(4) Fourth, the author should be allowed by the chosen university a free access to the course and to follow it through.

For the current study, to best meet the criteria of selection, the case used was an undergraduate course project initiated particularly for this research targeting juniors in Department of Translation and Interpreting in SIS, SYSU. The project was administered in the four-week Summer Term[1], lasting for 36 academic hours, which is the normal length for a course of 2 credits at the tertiary level in China. That is, evenly distributed across the four weeks, students met three times a week, namely on Monday, Wednesday, and Friday, with each time lasting 3 academic hours.

CAT has been an independent course in the department for 6 years which is offered in the "big" term, spanning 18 weeks with one meeting of 2 academic hours each week. It has always been taught in the traditional teacher-centred way given the lack of an ideal alternative teaching approach on the one hand, and the large class size (45-60 each class) on the other. The new approach was not implemented directly in the course seeing the great uncertainty in the sudden innovation. Yet a natural context was nevertheless desired to allow the study to produce natural findings. That is why the author initiated a course project in the "small" term, making it an elective yet formal course of 2 credits to students and hiding the real purpose of experimentation from them, so as to create a nearly natural context for this test and guarantee to a larger extent the reliability of the results.

31 candidates applied for this project who were later screened based on their resumes in which they were required to include at least their personal information, academic performance, and their prior experience in translation practice and use of TT as well as their interest in pursuing translation as their future career. The key information contained was purposefully prescribed to cover the 4 major factors the researcher would consider in selecting participants for this project, namely gender, academic performance, prior experience and their interest in pursuing translation as their future career. Diversity in the participants was the deliberate design in this

---

[1]   An academic year in SYSU consisted of three terms, starting with Summer Term of 4 weeks (normally between August and September), followed by Autumn Term (normally between September and January) and Spring Term (normally between February and June), both lasting 18 weeks. With a contrast in time length, Summer Term was often addressed by students as the "small" term while the other two terms the "big" terms.

selection, as inspired by Merriam (1998), stating that "findings from even a small sample of great diversity" produces key patterns that cut across cases and "derive their significance from having emerged out of heterogeneity" (p. 63).

Eventually, 25 students were chosen which could be grouped into 5 teams of 5 with different degrees of heterogeneity in the 4 factors. Specifically Team One features the most balanced gender structure (2 boys and 3 girls) while Team Five has all female members. Team Two is characterized by the sharpest contrast in the students' interest in translation as a profession, hence different degrees of motivation to learn. What makes Team Three distinguishable from the others is that the academic performances of the 5 team members are the most evenly distributed while, just on the contrary, those of the 5 ones in Team Four are almost on the same above-average level. This way of grouping allowed the author to observe the possible influence of the 4 factors on the learning process and outcomes of different individuals (see Appendix 1 for detailed information about the participants and their grouping).

Although all the participants were warned against dropping the course half way, a member of Team Five (i.e. Luo D.N.) told the researcher she could not make it to the course just a few days before the Summer Term began. Unable to recruit a new member, the researcher could do nothing but leave Team Five with four members only. She decided under the circumstances to observe the effect of the different group size on the learning process and outcomes, if any.

Besides the above distinguishing factors of each group, some prominent individual characters are also indicated in the table in case they have any bearings with the learning outcomes.

The author took the role of the tutor in this case since she was the best option given her rich experience in teaching CAT for 6 years and great familiarity with PBL. Although this was the first time she implemented a PBL course officially, the author had introduced and applied the idea of PBL from time to time in part of her teaching previously. Experience had accumulated to enable this authentic whole-sale try. Plus, consistent with the underlying notion of case study research, acting as a participant, the researcher could construct her knowledge from inside the story and understand better the proceeding of the case.

3. 3. 3. 3. 3   Data collection method

Once the case was bounded with questions to answer within certain time and place, the researcher had to plan the details of the study, determining how each question can be best answered (Lapan *et al.*, 2012). In other words, the following part will report what data she collected from whom by what means. For case study research, data collection and analysis were actually going hand in hand, iterative and interactive; yet for the convenience of narration, they are described here in separate sections.

1. The protocol and instruments for collecting data

It has been widely acknowledged that different sources of data are desired for case study research to increase the validity and trustworthiness of findings (Guion, Diehl & McDonald, 2011; Lapan *et al.*, 2012), which is termed *triangulation*. As Guion *et al.* (2011) summarised, there are at least five types of triangulation which are respectively data triangulation, investigator triangulation, theory triangulation, methodological triangulation and environmental triangulation. One thing to note is that it is not the mere goal of triangulation to achieve consistency across different data resources or approaches. Inconsistencies found thereof do not necessarily turn the evidence useless; rather they " should be viewed as an opportunity to uncover deeper meaning in the data " (Patton, 2002, cited from Guion *et al.*, 2011, p. 1).

To enhance the validity of the current research, the author adopted data triangulation and methodological triangulation with a hope that with multiple sources of data collected in different ways the author could have a more comprehensive view of the case and consequently gain a better understanding of the process.

More specifically, by data triangulation, the author collected data from different sources, which, in this case, include the learner participants, the teacher researcher and the classroom observer (see Figure 3. 2). In this way, the multiple perspectives could then provide a more comprehensive view of the phenomenon investigated.

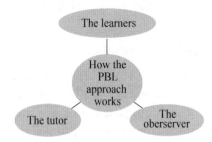

Figure 3. 2   The data triangulation of the case study

By adopting *methodological triangulation*, the author collected these data via different means or using different instruments as Table 3. 1 shows below.

From the learner participants, the author relied on a wide array of methods ranging from a questionnaire of open-ended questions, the final reflective reports, individual journals, group study records, classroom observations to their assessment documents. Reasons for adopting them as instruments of data collection will be elaborated below.

Table 3. 1   The methodological triangulation of the case study

The learners	The tutor	The observer
· Questionnair of open-ended questions · Individual journals · Group study records · Reflective reports · Assessment documents	· Tutor journals	· Obseration sheets · Notes of conversation with the researcher

First, the adoption of a questionnaire of open-ended questions. Interviews which are usually the best choice for such an evaluative study for instructional design were found infeasible for the current study because of the big size of the case on the one hand and the power gap between the researcher and the participants of the case study on the other. For one thing, the size of the class (i.e. 24 students) was too big to allow the researcher to process the enormous amount of data in anticipation within her time frame. For another, the researcher assumed the role of the tutor in this case study for the benefit of gaining some insider experience of the PBL approach from the tutor's perspective. This benefit, however, causes an inconvenience for the choice of interview as a way of data collection as being tutor and researcher at the same time, the author would pose great pressure on the learners during the face-to-face interview. In the cultural context of China, the power gap in between might prevent the learners/interviewees from expressing freely their opinions, especially those negative ones, mostly in fear of offending the tutor. Thus in compromise, the author followed the suit of some studies on early-stage PBL implementation (e.g. Macdonald, 2004) and devised a questionnaire of open-ended questions. This special survey in the place of the interview was employed in order to keep the advantage of allowing the participants some degree of freedom in expressing their opinions and at the same time avoid the direct face-to-face pressure. In designing the survey questions, Kirkpatrick's four-level model was referred to as it is admittedly the most widely used model in the field of training programme evaluation (Kirkpatrick, 1998). In light of the objectives of the current study, only the first three levels were adopted, namely reaction, learning and behavior. According to the model the questions at the first level sought the participants' *reaction* to the training, that is, the measurement of their satisfaction with the training. The questions at level 2 were concerned with *learning*, that is, to what extent the participants had changed their attitudes and increased their knowledge and/or skills as the result of the training. Then the questions at level 3 were intended to measure their *behaviour*, which inquired what change in the participants' behaviour had occurred because of their attending this course.

To effectively solicit information for evaluation of the three levels, nine questions were designed for the questionnaire survey with questions 1, 6, 7 and 8 addressing the first level, questions 2, 3 and 4 the second, and questions 5 and 9 the third (see Table 3.2). What is worth noting here is that the questionnaire was not delivered to the participants as a whole at once. The 9th question was reserved till one year after the course was over, as the author was aware that changes in the participants' behavior might not be noticeable shortly after the course was over. More changes were possible later when they did get opportunities to learn or practice more translation. That is why, both addressing *behaviour*, question 5 was delivered in the first part of the questionnaire to observe the immediate changes in the participants' behaviour while question 9 was reserved for one more year to capture those in the long term.

Table 3.2   The three-level questions in the questionnaire

Evaluation level	Question Number
1	1, 6, 7, 8
2	2, 3, 4
3	5, 9

The questionnaire was first piloted with three colleagues of the researcher who all had rich experience in teaching and researching. Suggestions were taken from them and revision made to the original questions in both wording and content. For example, the question of level 1 was originally phrased "How do you think of the effects of the PBL approach to this course?". All the three colleagues found "effects" here confusing and tended to lead respondents to answer with "what they had learnt from the course" which overlapped with the questions of level 2. Therefore the question was rephrased "How do you think of the new PBL approach to this CAT course? Please give brief reasons for what experience it has brought you, be it happy or painful." making it clear that what was asked about was their general feeling towards this instructional approach. Besides, a question of level 3 asked "Do you think this course with a PBL approach has caused you any changes in your learning habits?". It was suggested that with "in your learning habits", the question set an undesirable limit on the possible answers preventing respondents from showing other changes as a result of this learning experience. Thus in revision the phrase "in your learning habit" was dropped.

The final revised questions can be found in Appendix 2 and their distribution on the three levels is shown in Table 3. 2.

The next source of data is the reflective reports. The participants were asked to note down what impressed them most during the course and anything they wanted to ask or say to the tutor. They were allowed great freedom to decide what else they wanted to write and how much they wrote. These reports were expected to complement the above survey capturing something beyond what could be covered by the limited number of questions.

Journals of individual learners and group study records were used to provide evidence for the changes that might occur during the participants' learning experience, such as their knowledge acquisition, learning habits, attitudes. Journals, or diaries, have been advised by many researchers to be a good data source which "captures life as it is lived" (Bolger, Davies & Rafaeli, 2003, p. 579). By keeping track of the events and experiences in their natural and spontaneous context, journals are believed to help record "change process during major events and transitions" (ibid.) and reduce the distortion reality and are therefore more reliable evidence for what has happened (Clayton & Thorne, 2000).

For the similar reason, the summative assessment documents were taken as one of the data sources as they would show the researcher hard proof for the learning outcomes of the learners, which could help evidence or falsify the self-reported learning outcomes and changes in the survey and the reflective reports. Such documents constitute the artefacts in which the researcher can seek meaning in interpretive analysis (Charmaz, 2001; Glesne, 2006). The inclusion of all these data sources of different nature (i.e. some self-reported and some factual) were, therefore, expected to reveal more problems with a comparison between each other and consequently shed great light on the analysis results.

To complement the data from the learners' perspective, assuming the teacher researcher role, the author also kept journals herself after each meeting. Writing down her own thoughts about the research progress not only helped store accounts of the researcher's "observations, feelings, reactions, interpretations, reflections and explanations" (Elliott, 1991, cited from Donnelly, 2004) which provided the study with a researcher's perspective but also deepened the researcher's understanding of the instructional design, making herself more critical and cautious when making judgment. There were no fixed pattern for this journal, but the content was mainly centered on how the researcher found about her own teaching and how the learners changed as she perceived during the course. The tutor's journals were expected to verify or supplement the learners' self-reported findings.

Lastly, the author adopted a classroom observer, an advanced learner who had taken the CAT course in the "big" semester, the one taught in the transmissionist way. Out of the concern about the naturalness of the setting, the observer was designed to be sitting in the class as an auditor, keeping observation in secrecy since students were not aware of the research going on then. For the same reason, videotaping was not done as the researcher got to know that some participants felt very uncomfortable in the presence of a camera through casual conversations before the project. The observer was given an observation sheet to fill in, but was allowed freedom to make field notes and conversations with the students about anything related to the current study.

The advantage of setting an observer was two-fold: on the one hand, she could help the tutor note down the proceeding of the course in detail which would show the researcher whether what the participants said they did is the same as what they actually did in reality. On the other hand, as a learner who had been long used to the traditional way of teaching, the observer could easily sense the differences between PBL and the traditional approach, the effects of such a new approach on the old-fashioned learner, and, even more importantly, places where the tutor failed to execute PBL in the real sense.

2. The implementation

There are two things worth noting about the implementation procedure. The first one is the researcher's attitude. As Yin (1994) points out, there are no "routine formulas" for case studies, the researcher would have to "remain open to discovery of the unanticipated or unexpected" (Lapan *et al.*, 2012) although the plan had been set as shown above. What is found early in the study might inform or even improve how the remainder of the study proceeds. Therefore, during the data collection process, the author kept a curious attitude toward the case, trying her best to understand the case from the different perspectives prescribed and seeking for answers. The other thing is that out of ethical consideration, the purpose of this project was not revealed to the participants until the course was over and the researcher obtained *voluntary informed consent* as defined in the Belmont Report (National Commission, 1979) from all the participating students before the data and personal information were used in the current study.

The coming part introduces in detail how the data collection was implemented.

(1) The individual journals and group study records

The learners in this course were asked to keep two kinds of journals: one on individual basis which had to be kept every day and submitted via email to the researcher before 10 p.m. every night, namely the *individual journals* and the other on collective basis by teams, which had to be kept each time the team met and be submitted via email, too, to the researcher every time it was done, namely the *group journals*. As most of the students had not had the habit of keeping journals about their study, the researcher designed a template for the individual and group journal (see Appendix 3 and Appendix 4) respectively for their reference, but they were allowed to use other forms of journals as well. All the journals were kept in the electronic form and fed into QSR NVivo 10 for analysis.

The individual journals were collected with the "time-based design" with a fixed interval (Bolger *et al.*, 2003, p. 588). The learners were asked to keep journals every day, including those days when they did not have classes. This decision was made based on the following consideration: the PBL approach requires a substantial amount of self-study outside the classroom. What the learners did on those days with no class are even more important indicators of their progress. The deadline for the submission of the journals was set on

10 p.m. every day given that the participants enrolled also in other courses and engaged in other activities during day time and thus most of them could not sit and reflect on their learning till evening. Therefore, the author made the time at 10 p.m. to allow the learners sufficient time to keep full record of their learning experience of the day and also leave herself enough time for a scanning of the journals, keeping abreast of their progress which might in turn inform necessary adjustment.

Besides, as suggested by some researchers (e.g., Bolger *et al.*, 1989; Laurenceau, *et al.*, 2005), to make the participants open their hearts and keep truthful about their feelings and personal stands, promises had been given to assure them that their journals would be kept classified and nobody but the researcher would see them.

（2）Tutor journals

As suggested by many researchers (e.g. Berg, 2007; Janesick, 1998), the author kept journals, too, in which she noted down her behavior and reflections during the tutoring process in and outside the classroom. Journal writing is believed to enjoy a two-fold benefit: such documentation may help deepen the researcher's understanding of the study and the saliently relevant written thoughts can be incorporated into the writing of the final report making our research more public. As Janesick (1998) states, journal writing is "a type of connoisseurship by which individuals become connoisseurs of their own thinking and reflection patterns and indeed their own understanding of their work" (p. 24).

The author shares the idea that journal writing is "a tangible way to evaluate our experience, improve and clarify one's thinking, and finally become a better … scholor" (ibid., p. 3).

The tutor journals all started with a brief note on the context in which it was written. The content was not kept the same across the journals of different days. Emerging thoughts were noted down and the key aspects of them recurrent during the study were highlighted with headings for the convenience of later comparison and analysis. The journals were sequenced chronologically and also fed into QSR NVivo 10 for later analysis.

(3) Observational sheets

This research adopted an advanced learner as an observer. Since the observer had no prior experience, the author organized an observer training for her. In the training the oberserver was asked to observe 3 other classes the author taught with exact the same procedures as in the current study. Discussions were held each time after the observation until the observer believed she had been ready for the real one.

Since the participating students were not aware that this course was experimental, videotaping was not possible for the observation. Thus only direct observation was done. To facilitate the observation, a classroom observational sheet was designed (See Appendix 5) with a structure indicating what to note down. There were two reasons for the offering of this observational sheet to the observer: on the one hand, with the observer lacking experience in classroom observation and under the circumstances where no videotaping was possible and thus all the pressure was on her shoulder, the provision of such a sheet could free her from the burden to make on-the-spot decisions about what to note down which would inevitably lower the quality of her notes in terms of both completeness and accuracy. On the other hand, with the guidance of the sheet, the observer's notes across different meetings would be much better structured and comparable.

Meanwhile, the observer was given freedom to record what she found interesting or worthy of noting beyond the protocol. In addition, the observer was encouraged to exchange with the tutor during and after the observation her feelings and questions. Key points in the conversations were kept down as field notes for later use.

In sum, data from the observer included the records on the classroom observational sheets and the short conversations the researcher had with the observer during or after the classes. This part of data were expected again to provide hard evidence for the proceeding of the classroom teaching and the possible changes in the learners' behaviour in response to the teaching.

(4) Assessment documents

The assessment documents used as data sources include the

reflective report and the documents in the final group-based computer-aided translation project.

At the end of the course, as part of the final assessment, the learners were required to hand in a reflective report individually, summarizing what they had learnt from the course, how they commented on the approach to the course and asking whatever questions they still had about the course. Students were encouraged to be frank when giving their opinions, be they positive or negative, and say whatever they wanted to say to me beyond the guiding instruction. The reflective report was included as part of the final assessment considering that the learners might be feeling tired at the end of the course and simply brush it off if it was not related to the final grading. Yet the author was aware that under the pressure to win a better mark, however, the learners would possibly fill their report with stuff of which the negative opinions were deliberately reduced with a conscious or unconscious intention to please the tutor. Special attention was therefore paid to the content of the reports and those without any negative feedback would be anlysed with caution.

Table 3. 3　Summative assessment documents for data analysis in alignment with intended learning outcomes（ILOs）

Assessment Documents			Objects of assessment
Project plan (first & final drafts)	Project description		ILOs 6, 7, 9
	Workload analysis		
	Work division		
	Work flow		
	Time management		
	Risk management		
	Quality assurance measures		
	Technological aids	What	ILOs 1, 2, 4, 7, 9
		Where	
		For what purpose	
	Miscellaneous		ILO 12

Continued

Assessment Documents		Objects of assessment
Translation project process management documents	Translator training	ILOs 1, 2, 3, 4, 6
	Team member communication/meetings	ILO 10
	Process monitoring	ILO 6
	Personnel evaluation	ILO 11
	Process documents out of translating on CAT	ILOs 3, 5
	Communication with client	ILO 6
	Miscellaneous	ILO 12
Quality assurance (QA) documents	Style guide	ILOs 6, 9
	Translation brief	
	Translation quality evaluation criteria	
	QA process documents	
	Translation error collection	
	Miscellaneous	
Final delivery	Translated texts	ILO 12
	Translation memory & term base maintained	ILOs 3, 5
Summative reports	Reflections on the translation project management	ILOs 9, 11
	Reflections on the choice of CAT tools	ILO 7
	Evaluation of the adopted CAT tools	ILO 8
	Individual reflective reports	ILO 12
Experience exchange		ILOs 6, 7, 8, 9, 12

The other part of the assessment documents were those produced during the final project as a summative assessment. What were used as data sources are displayed in Table 3.3 above. On the right column of the table is the intended learning outcomes (ILOs) the documents were intended to assess.

(5) The survey of open-ended questions

Since the participants were not aware of the research going on, the survey was carried out after the course was over. Before responding to the questions, the participants were told that the grading for the course had been done so that they would feel safe to express freely their own opinions. Besides, the researcher also revealed to them the intention and the significance of this study to encourage the participants to be frank with themselves, freeing them from the concern of displeasing the tutor. The first part of the survey was done with the researcher in presence in order for the respondents to feel proper pressure to be serious about it and guarantee the reliability of the responses. 24 questionnaires were delivered and were all returned with all the questions answered although with different degrees of depth. The 9th question was delivered to the 24 participants via email one year after and received 21 returns.

To sum up, the author adopted 7 data sources ranging from learner journals to the assessment documents triangulated in collection instruments and data sources. Considering the adaptive and dynamic nature of case study research, the author decided to collect data on daily basis so that she could process them in time and adapt her research if necessary. The only exception was the survey. As explained earlier, the purpose of the course project was kept a secret to the students before it was over. The survey could, therefore, not be carried out until the course was finished altogether.

3.3.3.3.4 *Data storage and management*

Electronic tools for general and specialised purposes were employed for the data storage and management.

Firstly, all the data were produced or transformed into electronic forms for the convenience of storage and exchange. To be more specific, all the journals, reports, survey and academic performances were produced in the electronic form; the observation sheet and conversations were firstly hand-written and later transformed into the electronic form.

Then, *Dropbox*, a web-based digital storage service, was adopted for storing, sharing and synchronization of data. More exactly, each participant was demanded to apply for a *Dropbox* account, and install the client terminal on his/her PC or laptop which had to be internet-connected. In this way, when a new folder was created on the website and shared by all the participants, it would appear automatically at a specified location on their computers. The participants would then enjoy access to the folder via their own terminals, and any changes any one of them made to the folder (e.g. adding a file, editing an existing file, deleting a file) on his/her own client terminal could be synchronized on all the others' computers as well as in the original folder online. Storing, sharing of files and coordinated work on the same file were thus able to be accomplished at the same time and much more easily. To avoid the chaos as an inevitable result of too many users having access to the single folder, a structure of storage (Appendix 6) was created and an instruction of *How to use Dropbox* (Appendix 7) issued to keep the files organized and safe.

To reduce the risk of data loss to the minimum, a copy of all the data obtained was instantly made and stored on *Baidu Cloud*, a cloud storage service provider in China. In sum these syncing and storage tools enabled the researcher to easily obtain, store and organize her data.

Lastly, given such common features of qualitative research as "subjectivity, richness, and comprehensive text-based information" (Hilal & Alabri, 2013, p. 181), the current research would inevitably anticipate a large amount of data from multiple sources. An analysis in search of the relationship between categories and themes of the data was expected in turn to lead to a fuller understanding of the case in study. Seeing that such analysis is often "a muddled, vague and time-consuming process", innovative technology has been designed to simplify the difficult job, among which the software NVivo has been regarded as the best of its kind. Therefore, its latest version

QSR NVivo 10 was employed to manage the coding procedures for this study. The procedures of data management and analysis the author followed were based on the suggestion by Hilal & Alabri (2013, p. 182) adapted from Bazeley (2007).

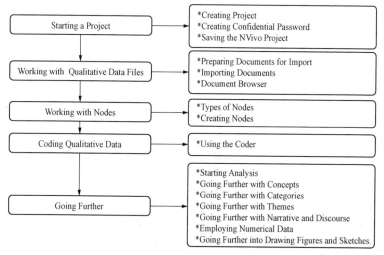

Figure 3. 3　Procedures followed in applying QSR NVivo Software (Hilal & Alabri, 2013, p. 182)

3. 3. 3. 3. 5　Data analysis method

This section is aimed to introduce the analytic methods adopted to bring order and meaning to the mass of data collected. The data analysis actually started in parallel with data collection. The author was aware that with data collected on daily basis, data collection and analysis technically should better take place "simultaneously in a dynamic and interactive" way (Lapan et al., 2012, p. 263), with major issues to follow emerging gradually. Therefore field notes and dated memos were written every day especially in early days of the processing stage, noting down "hunches" or primitive analysis that might inform later interpretation.

Systematic analysis was carried out upon the completion of the data collection process. There were three steps in the analysis phase, namely data reduction, data organisation and data interpretation.

1. Data reduction

First, as elaborated above, the broad range of data sources in this stage of study produced inevitably a large amount of raw data. The author therefore would have to decide the way of data production so as to reduce the massive data into a manageable form which allows for systematic and focused analysis and enables the researcher to understand better the phenomenon under study. As Miles and Huberman (1994, p. 11) argue,

> Data reduction is not something separate from analysis. It is *part* of analysis. The researcher's decisions — which data chunks to code and which to pull out, which evolving story to tell — *are all analytic choices*. Data reduction is a form of analysis that sharpens, sorts, focuses, discards, and organizes data in such a way that "final" conclusions can be drawn and verified.

So as part of data analysis, a data reduction plan of three steps was worked out in light of the research questions of this stage of study, namely to find out how the participants perceived the CAT course with the PBL approach. The first step was to delineate the boundary of data sets, or more exactly, make clear what data sources would be coded for analysis and in what order. For the current study, the data sources to be coded were decided to cover the individual and collective journals of the learners, the learners' reflective journals, the teacher journals, the learners' assessment documents, and the observation sheets and field notes of the informal conversations during observation. The above mentioned sources neatly fall into two categories — self-reported data and factual data. Then it was decided that self-reported data were to be coded first. After it the factual data were analysed to provide further evidence or reveal counterevidence for the findings among the self-reported data so that more justified conclusions could be made. In sum, the above mentioned raw data were analysed in the order specified in Figure 3. 4.

Each of these data sources was analysed separately and then integrated using the common coding system as defined in Table 3. 4.

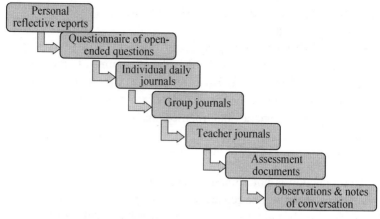

Figure 3.4   The order of raw data analysis

Table 3.4   Codes for themes for theory-driven analysis of the raw data

Kirkpatrick Evaluation Level	No.	Codes for Themes	Knowledge type	ILOs
Reaction	1	Satisfaction with the PBL approach	-	-
Learning	2	Knowledge about CAT, MT and translation project	Declarative knowledge	ILOs 1, 2, 4, 6
	3	Knowledge of CAT technologies operation	Declarative knowledge	ILO 3
	4	Attitude towards CAT and CAT learning	-	-
Behaviour	5	Ability to apply CAT technologies in practice	Functioning knowledge	ILOs 5, 7, 9
	6	Ability to monitor and assess one's performance in applying the learnt knowledge in a translation project	Functioning knowledge	ILO 11
	7	Ability to reflect upon one's performance with CAT technologies and evaluate functioning of the technologies	Functioning knowledge	ILO 8

Continued

Kirkpatrick Evaluation Level	No.	Codes for Themes	Knowledge type	ILOs
Behaviour	8	Ability to cooperate with others	Functioning knowledge	ILO 10
	9	Ability to learn independently and continuously (including motivation and problem solving skills)	Functioning knowledge	ILO 12

Then the second step was to decide the exact data analysis approach. In this case the theory-driven theme-based approach proposed by Namey *et al.* (2007) was adopted. According to Namey *et al.* (ibid.) thematic analysis of raw data involves codes developed for identified themes which are then applied, as "summary markers" (p. 138), to later analysis of raw data by comparing frequencies of themes, looking for codes co-occurrences and so on. Of the two approaches to qualitative data analysis, namely the data-driven analysis and theory-driven analysis, the theory-driven approach was adopted as Namey *et al.* (ibid., p. 139) points out that it is comparatively more "structured" and therefore "may be considered more reliable" especially when there is only one coder in this case.

The third step then was to work out a series of codes for themes that would guide the actual analysis later. To strike a balance between reliability and validity of the data analysis, the author took into consideration the following two aspects of information when determining the codes. One of them is the open-ended survey questions. "The discrete questions and their relevant probes" provide the author with a set of "structural codes" as defined by Namey *et al.* (2007, p. 140). The other aspect is the data in themselves. As Namey *et al.* (2007) points out, the theory-driven approach does not exclude the data-driven approach altogether. Instead sometimes both can be alternated to achieve a better balance of validity and reliability. The author therefore read closely the participants' reflective reports and developed codes for the emergent themes identified by using a data-driven approach (see the resulting codes in

Table 3. 4). It is hoped that in this way no useful information in the raw data would be missing due to the subjectivity in the researcher's predetermined theory a priori.

The codes for later analysis were finalized by incorporating the codes developed from the above two routes within the evaluation framework of 3 levels adapted from Kirkpatrick (1998). They are presented below in alignment with the types of knowledge and the ILOs of the course, in which way the findings from the data could be correlated with the learning outcomes easily, hopefully allowing for a more justified judgment about the effects of the design in the end.

To make the analysis more feasible, the verbs for ILOs from Bloom's revised taxonomy (see Tables 5. 2 and 5. 3 in Section 5. 1. 2. 2. 1) were used as indicators to help identify information in the raw data evidencing the themes in the table above (Table 3. 4).

2. Data organisation

Following the reduction of data is the second step, data organization. All the data were transformed into electronic texts and fed into QSR NVivo 10. Then adopting the theory-driven approach, the researcher read the text as many times as necessary and coded the data with the codes while at the same time keeping an eye on recurring regularities in the data that might have evaded the original theorization.

3. Data interpretation

The last step in data analysis was interpretation. Decisions and conclusions would be made in response to the research questions. Details of this step will be reported in next chapter.

3. 4　A summary

In this chapter, the author elaborates the research design in detail. In light of the research objectives, the current study was defined as instructional design research across *exploratory* and *design & development* stages. Of this research type, the current study falls naturally into the *interpretive* paradigm with the major goal of understanding how the new approach works with participants rather than making judgments of causality or superiority. Procedures were decided then with reference to R&D which were further categorized into three stages addressing more specific research questions broken down from the objectives. An overview of the research design is shown in Figure 3.5.

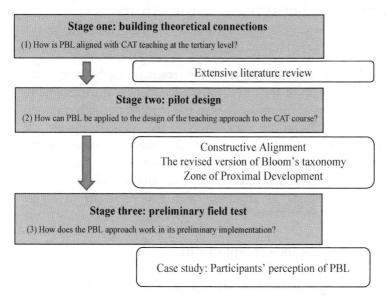

Figure 3.5   An overview of the research design

Chapter Four
Building Theoretical Connections

In the previous chapter, the research methodology was elaborated in light of the three research questions that correspond to the three stages of the study. Now in this chapter, the author aims to report the results of the first stage of the research, namely the theoretical connections between PBL and the CAT course at the tertiary level. Drawing on the previous studies, this chapter will examine the concept and theoretical foundations of PBL on the basis of which to decide its alignment with CAT teaching in higher education.

4.1 Clarifying the concept of PBL

With the widespread use of PBL at different levels across the world, there has, almost inevitably, appeared various practices and models due to the different educational systems and cultures where it is implemented (Kolmos, 2009). The implication of this fact is two-folded: first, different definitions and models of PBL have to be distinguished carefully before one can be found to be relevant to the aims of the current study; second, it is impossible to find a ready-to-use model for a PBL approach to CAT teaching in tertiary translation programmes particularly. The lack of a precise understanding of PBL may lead to misapplications of it and as a result may fail to attain expected outcomes without knowing why (Boud & Feletti, 1997; Maudsley, 1999; Savery, 2006). Clarification of the concept is, therefore, highly needed here before necessary tailoring or creation can be made to fit a candidate model into the CAT teaching context.

The author therefore first identified the possible sources of confusion about PBL and explicate the relationship between it and a few other student-centred instructional methods that are easily confused by people. Then the definition of the PBL was worked out on which the design framework was based on.

4. 1. 1   Identifying sources of confusion about PBL

As mentioned above, PBL exists in many forms which are various depending on the perspectives of the users or the contexts of application (e.g. level of education, discipline, culture). A simple search on Google showed us an amazing diversity of theories and practices covered under the common label PBL. What PBL denotes range from a philosophical learning theory to various educational practices. It should be mentioned here that what the author intended to do was identify the core version of PBL to be used in this study without challenging any of those claims to PBL.

The first source of confusion arises out of inconsistency in what is the nature of PBL. To illustrate it, a few definitions are quoted here:

(1) The very original definition was made by Barrows and Kelson (1995) who defined PBL as both a curriculum and a process. The curriculum consists of carefully selected and designed problems that demand from the learner acquisition of critical knowledge, problem solving proficiency, self-directed learning strategies, and team participation skills. The process replicates the commonly used systemic approach to resolving problems or meeting challenges that are encountered in life and career.

(2) Finkle and Torp (1995) defined PBL as " a curriculum development and instructional system that simultaneously develops both problem solving strategies and disciplinary knowledge bases and skills by placing students in the active role of problem-solvers confronted with an ill-structured problem that mirrors real-world problems" (p. 1).

(3) Samford (2003) defined PBL as an instructional strategy that promotes active learning. PBL can be used as a framework for modules, courses, programs, or curriculums.

(4) Savery (2006) defined PBL as an instructional (and curricular) learner-centered approach that empowers learners to conduct research, integrate theory and practice, and apply knowledge and skills to developing a viable solution to a defined problem.

All in all, as Graaff and Kolmos (2003) pointed out, PBL can be used to "refer to theory, models, and practice" (p. 658). In addition, when used as a theory to guide teaching practice, we can see in the above definitions, it can be and has indeed been applied to levels ranging from a single course to a whole programme (Taylor & Miflin, 2008).

Secondly, different models of PBL can be distinguished by the way they are implemented. The two most famous models of PBL are the problem-based learning model practiced mainly in health sciences at McMaster University and the problem- and project-based learning model practiced at Aalborg and Roskilde University in a greater range of subject areas such as engineering, science, social science, and the humanities. Apart from the different scopes of implementation, the biggest difference between these two models is the involvement of teacher direction. Comparatively speaking, students work more on the cases defined by teacher in the problem-based model while problems in the problem-and project-based model are more identified by students themselves and solved in the form of projects with common products worked out collectively (Kolmos, 2009, pp. 264-267). Seeing the quick increase in the variations of PBL models, Barrows (1986), Harden and Davis (1998) and Hmelo-Silver (2004) have attempted to categorize PBL models. Hung (2011) suggested that Barrows' taxonomy along the two variables of self-directedness and problem structuredness be used as a structural framework to differentiate different models making possible a comparison between them as well as between their effectiveness.

The third distinction is based on the different purposes of the

PBL model. Savin-Baden (2000) differentiated five purposes of PBL implementation which are respectively PBL for epistemological competence, PBL for professional action, PBL for interdisciplinary understanding, PBL for trans-disciplinary learning, and PBL for critical contestability.

Adding to the confusion in the concepts or models of PBL is the use of the name to cover some variations that are far from, if not running counter to, the original concept. For example, PBL is described in some older medical schools as "an adjunct to more traditional lecture and laboratory-based instruction" (Steele, Medder & Turner, 2000, p. 23) or a combination of problem-based learning and information-based learning (Harden & Davies, 1998). Some version of PBL even reported to use students as tutors because "cases are designed primarily to reinforce and to supplement information presented in lectures and to provide students with opportunities to use their knowledge to solve clinical problems" (Steele *et al.*, 2000, p. 24).

In addition to the internal confusion in different modes of PBL under the common term, PBL is very often confounded with other student-centred approaches such as project-based learning, discovery learning, case study method and problem-solving method.

The first pair of concepts to differentiate is project-based learning and problem-based learning which have both been shortened as PBL and are easily confused not only in form but also in semantic content. The confusion arises mainly from the different definitions of *project-based learning.* Among the well-known ones is one mentioned above that treats *project-based learning* as synonymous with problem-based learning holding that it shares the basic principles with *problem-based learning* while differs only in the format of problems and the degree of teacher directions, constituting one of the models of *problem-based learning.* Yet there are some researchers who maintain that there exists a substantial difference between *problem-based learning* and *project-based learning* as they find *project-based learning* synonymous to *task-based learning* with projects defined as narrowly formulated task instead of an open learning process (Prince & Felder, 2006; Savin-Baden, 2003). Some other scholars even make *project-based learning* a generic term subsuming *problem-based learning* by defining projects as "a complex, unique and situated task that will always involve an open approach" (Algreen-Ussing & Fruensgaard, 1990, cited from Kolmos, 2009, p. 268).

PBL differs significantly from *discovery learning* regardless of their common advocation of *learning by doing* by stressing the importance of open-ended problem-solving for students' construction of knowledge. The biggest distinguishing feature is the way the problem is given and solved in the two approaches: in PBL, students, without any preceding teaching, are given problems to solve by integrating into their prior knowledge their new acquisitions via communication and research collaboratively within and outside the problem with exploration of external resources greatly encouraged. Learning is expected to take place during the socially based problem-solving process. Differently in *discovery learning*, necessary pre-teaching is provided (especially in enhanced models) to assist students with their learning that is supposed to take place when tackling the given problems through working with their peers to discover new perspectives but without reference to external knowledge. As Alfieri *et al.* (2011) found in their review, "the target information must be discovered by the learner within the confines of the task and its material" (p. 2). Moreover teamwork is not as important in discovery learning as it is in PBL where it is among the must components.

The other two approaches that have very often been mentioned in parallel with PBL are the case study method and the problem-solving method. What is behind their deceiving similarities on the surface is the enormous divergence in their notion of knowledge. While PBL conceives knowledge as being constructed by learners conquering problems when communicating with peers in the social context, the case study method and the problem-solving method fall in the traditional paradigm of education considering knowledge as being transferred from teachers to students. Cases and problems are used in these methods after knowledge has been transmitted to students for the purpose of enhancing students' analytical abilities in applying the newly acquired knowledge to a problem or a case. As Ross (1991) highlighted, in problem-based curriculums, knowledge arises from working on a problem rather than, as with problem-solving, being a prerequisite for working on a problem. Therefore, in these two approaches teachers are in control of the learning process by guiding students through analysis to get the desired solutions expected by the teacher.

As revealed above, it is a plain fact that PBL has been extraordinarily diffused across a wide range of academic disciplines and transcending different levels of education. As a result, the differences and even confusion in understanding the innovative educational idea are inevitable (Rogers, 1995). The variations are so extensive and broadly grounded in the different contexts they are applied to, that Taylor and Miflin (2008) found any attempt to "rescue the term PBL" (e.g. Maudsley, 1999) by prescribing the common ground "doomed to failure" (p. 752). Instead, they believed that critical awareness of the differences and sources of them may help ease the frustration of those seeking to adopt or currently using PBL.

4. 1. 2   Defining PBL for the current study

Regardless of their gloomy evaluation of the current scene pertaining to PBL research and practice, Taylor and Miflin (2008) nevertheless admitted that certain features of PBL arising from analysis of its origins and evolution comply with the initial promise of this approach and are more conducive to achieving the goals of modern professional education, thus deserving "consideration of present and future practitioners of PBL" (p. 753).

Meanwhile we can find many reported efforts to identify the defining characteristics of PBL that underpin this methodology (e.g. Boud & Feletti, 1997; Charlin *et al.*, 1998, Duch, Groh & Allen, 2001b; Hmelo-Silver, 2004; Kolmos, 2009; Savery, 2006; Torp & Sage, 2002) with which to guarantee the consistency among different models or applications of PBL while allowing for flexibility and diversity unavoidable in concrete practices. This section then will be an attempt at a meta-analysis of these studies in search for an understanding to be adopted for the current study.

4. 1. 2. 1　Previous endeavours

Barrow, who is widely recognized as the most important figure who introduced PBL to life, has long before noticed the imprecision of the definition of PBL (Barrow, 1986, 1996) and expressed the need for a core model or basic definition to make possible the comparability of different implementations under the same framework (Barrow, 1996). Yet instead of providing a brief definition, he summarized six characteristics of the original model he had developed at McMaster University which were found fundamental to the PBL approach. They are as follows:

(1) learning is student-centred;
(2) learning occurs in small student groups;
(3) teachers are facilitators or guides;
(4) problem form the organizing focus and stimulus for learning;
(5) problems are a vehicle for the development of clinical problem-solving skills; and
(6) new information is acquired through self-directed learning.

Similarly Charlin *et al.* (1998) found that PBL had been too extensively diffused to be circumscribed within a brief definition. A more feasible alternative approach as they suggested is "to specify what most educators believe constitute its core, distinguishing principles" (p. 324). In this vein, Charlin *et al.* (ibid.) identified three core principles which are:

(1) the starting point of learning should be a problem which may be of different natures and in different formats;
(2) PBL is an educational approach rather than a specific teaching technique that is used sporadically in an otherwise traditional environment;
(3) PBL is a student-centred approach which should engage students in active processing of information, activate their prior knowledge, and provide students

with meaningful context in which to learn and opportunities to elaborate/organize their knowledge.

Furthermore, Charlin *et al.* (ibid.) identified ten discernible dimensions of the PBL approach based on the previous definitions and models with an aim to formulate a framework that is believed to be able to assist PBL planners with their decision making concerning how to use PBL in a given setting and facilitate comparisons between different implementations as well as judgments on the validity and reliability of reported effectiveness of PBL implementations. The ten dimensions are:

(1) the person or group who selects the problem;
(2) the purpose of the problem;
(3) nature of the educational objectives and control over their selection;
(4) the nature of the task;
(5) the presentation of the problem;
(6) format of the problem;
(7) the processes students follow;
(8) resources utilized and how they are identified;
(9) the role of the tutor;
(10) demonstration of learning through a product or a performance (assessing students' achievement).

So according to Charlin *et al.* (ibid.) implementations with adherence to the three core principles may vary from one another on the ten dimensions without threatening their status as being of the PBL genus.

Having witnessed the heterogeneous and even " subversive " activities going under the same term of PBL, Maudsley (1999) also made an attempt to disperse the "conceptual fog" and rescue PBL as a legitimate term. In conclusion he proposed the following *ground rules* that PBL:

(1) is both method and philosophy, curriculum-wide, and supported by all curricular elements;
(2) aims at efficient acquisition and structuring of

knowledge arising out of working through (in an active, iterative, and self-directed way) a progressive framework of problems providing context, relevance, and motivation (problem-first learning);

(3) builds on prior knowledge, integration, critical thinking, reflection on learning, and enjoyment;

(4) achieves its goals via facilitated small-group work and independent study; and possibly

(5) relates to problem solving only insofar as knowledge becomes more accessible, and can therefore be applied more efficiently, during this process.

Savery (2006) is another attempt to clarify the definition and unify understanding of PBL as an instructional approach. He reiterated the following nine "generic essentials" of PBL proposed by Barrows:

(1) Students must have the responsibility for their own learning;

(2) The problem simulations used in PBL must be ill-structured and allow for free inquiry;

(3) Learning should be integrated from a wide range of disciplines or subjects;

(4) Collaboration is essential;

(5) What students learn during their self-directed learning must be applied back to the problem with reanalysis and resolution;

(6) A closing analysis of what has been learned from work with the problem and a discussion of what concepts and principles have been learned are essential;

(7) Self and peer assessment should be carried out at the completion of each problem and at the end of every curricular unit;

(8) The activities carried out in PBL must be those valued in the real world;

(9) Student examinations must measure student progress towards the goals of PBL.

Then he reduced the nine characteristics into three foci of consideration in instructional design of PBL, namely *the role of the tutor, the responsibilities of the learners,* and *the essential elements in the design of ill-structured problems.*

Taylor and Miflin (2008) initiated a proposal for PBL in 2008 in the context of medical education at curriculum level. Nevertheless, the essential context-independent ideas can be safely extracted for applications in other disciplines and across different levels.

First, PBL should serve for the three interrelated and interdependent goals of education in the 21st century, namely to graduate professionals who are able to acquire, apply and update effectively and efficiently knowledge and skills as necessary to deal with new problems.

Second, self-directed learning should not be misconceived to equal unstructured and totally independent learning throughout the process; instead as original practised by Barrows, teaching should be carefully planned and structured in service of the goal of making students self-directed learners. Many researchers (Miflin & Price, 2000; Miflin, 2004; Taylor & Miflin, 2008) have also pointed out the confusion of self-direction for lifelong learning as a goal (product) with the entire learning process by which it is achieved.

Following this is the notes about problems. Two features are found critical to any PBL variation. One is that the problem comes first in the tutorial process. The other is that problems for PBL must be those that are prevalent and important in practice. Further, it is reemphasized that problems should be designed to encourage "whole learning" of all the necessary knowledge from different domains for solving them. Besides, they should be presented in such a way that problem solving stages learning as such rather than serves as "convenient pegs" to hang knowledge acquisition for later use which falls back to the traditional paradigm of "knowledge first, application later" (Taylor & Miflin, 2008).

In addition, Taylor & Miflin (ibid.) is among the very few publications which discuss in depth the sequencing of the content in PBL. In sum, it was suggested that all domains of knowledge, skills and attributes be introduced in both horizontally and vertically integrated way. The problem can neither be totally unfamiliar nor a mere rehash of an old one. It needs to be consistent with the stage of student learning (Davis & Harden, 1999) with required new learning to be within the *zone of proximal development* (ZPD) (Vygotsky, 1978) and structures students' learning in a logical way.

In the author's research, Barret (2005) seems to be the most inclusive try. Drawing on the previous studies of defining PBL and keep in line with the classical definition of PBL by Barrows, the well-recognized initiator of PBL in higher education, Barret (ibid.) defines PBL as a total approach to education consisting of four important components: PBL curriculum design, PBL tutorials, PBL compatible assessments and philosophical principles underpinning PBL (as shown in Figure 4. 1).

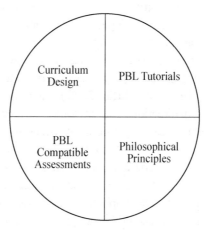

Figure 4. 1   PBL as a total educational approach (**Barret**, **2005**, **p. 15**)

Besides this theoretical model for PBL, Barret (2005) provided pertinent guidance on its operation. For example, the operational definition of "PBL as a learning process" (p. 15) can be readily used to guide the tutorials; the PBL "curriculum matrix" (p. 17) is highly helpful in a PBL curriculum design, ensuring the alignment between the problems and the expected learning outcomes; the "checklist of problem formats" (p. 17) provide PBL designers a variety of selections for their problems and "the practical advice on ways to be an effective PBL tutor" (p. 19) gives teachers of PBL a benchmark to evaluate their performance against.

4. 1. 2. 2   A synthesized view

Based on the review of the above studies, the author is highly aware of the commonalities as well as differences between PBL and other student-centred approaches to learning. Recognizing the impossibility to encapsulate PBL into a brief definition, the author decided, therefore, to describe PBL in the following way in order to grasp its gist without losing the whole picture. First the author will adopt Barret's model (2005) as a defining framework for PBL. Following it, she will demarcate the level PBL is applied to in this study in accordance with the research questions. After this are the very distinguishing features of PBL that separate it from other related concepts. Then the ground rules of a PBL design ensues before the educational goals of PBL approach are enumerated at the end. Explication of learning outcomes are indispensable for PBL designers not only because they serve as the benchmark for a measure of the PBL's applicability in various contexts but also because the design/selection of both the problems and the assessment methods have to be congruent with them.

4. 1. 2. 2. 1   *Understanding of PBL*

Barret's four-component model of PBL (Barret, 2005) is considered the most comprehensive and adequate framework of PBL. It shares with many other researchers (Hoffman & Ritchie, 1997; Uden & Beaumont, 2006) the understanding of PBL as a total educational approach. Yet with the four components specified explicitly, Barret makes his model more readily applicable (see Figure 4. 1). The author will therefore adopt the model as a framework of understanding PBL and its later design.

But as the current study is an initial attempt at introducing PBL into translator training programme in higher education in China, the author, with limitations in time and capability, would have it tried only on a single course — CAT in this case. Therefore, one of the four components, namely curriculum design, in the original model of

Barret will be replaced by course design for the current study. In other words, the author will apply the model to the teaching of the CAT course or employ the model as an instructional approach to a course teaching. In light of this understanding, PBL will in the following section be defined and modeled with a view to guide *how* the teaching and learning will be conducted in the course. In addition, the author finds that philosophical principles is not a component in parallel with the other three ones; rather it is one that governs the production and working of the other three. As the result of the analysis, the understanding of PBL by Barret (2005) was revised for this study as shown below in Figure 4.2.

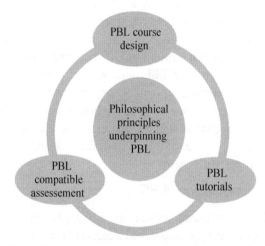

Figure 4.2　Framework of PBL adopted in this study
(revised from Barret 2005, p.15)

4.1.2.2.2　*Defining features of PBL*

As revealed in the above discussions, two factors emerge to distinguish PBL from other student-centred approaches, which are *the functions of problems* and *the role of teachers*. It is only in PBL that learning starts with problems instead of any form of teacher-dominated instruction such as lectures. In other words, no teaching is offered prior to the problem. Problems serve in PBL not only to stimulate learning but also to be a vehicle by which learning takes place. Knowledge is not acquired before the problem is given but during the process the problem is solved. In sum, problems occupy a central role in PBL for they stimulate, organize and provide meaningful contexts for learning. The second unique feature of PBL is the retreat of teachers from the foreground dominating authority to the background facilitating resources, although teacher interference may vary in degree among different models of PBL.

4. 1. 2. 2. 3   Ground rules of PBL design

Furthermore, in accordance with the above modeling of PBL, the ten dimensions of PBL proposed by Charlin *et al.* (1998), upon analysis, can neatly fall into three categories, namely problems, tutorials and assessment.

In designing the above three elements of PBL, some ground rules are found as necessary for the approach to work successfully towards the expected outcomes it purports to turn out. They are:

(1) Learning in PBL is problem-driven;
(2) Problems should be authentic and ill-structured as they are valued in the real world;
(3) Problems should be able to activate students' prior knowledge and provide meaningful context, relevance and motivation for learning to take place;
(4) Students are active participants who are responsible for and in control of their own learning;
(5) Students work both independently and collaboratively in (small) groups in the learning process with facilitators providing necessary scaffolding and guidance;
(6) The learning process should be a self-regulated iterative process starting with problems, building on prior knowledge and skills;
(7) The learning process should include synthesis, integration of and reflection on learning and end with evaluation and review of the learner's experience;
(8) Assessment should be well-aligned with learning outcomes and compatible with the PBL process.

4. 1. 2. 2. 4　Educational goals of PBL

Lastly, five educational goals of the PBL approach widely acknowledged (Barrows & Kelson, 1995; Hmelo-Silver, 2004) was adopted by the current study. As Hmelo-Silver (ibid.) put it, PBL is designed to help students

(1) construct an extensive and flexible knowledge base;
(2) develop effective problem-solving skills;
(3) develop self-directed, lifelong learning skills;
(4) become effective collaborators; and
(5) become intrinsically motivated to learn.

In sum, in this section the author depicted the core version of PBL with reference to its defining features, instructional planning elements, ground rules in design and educational goals. The next section will then be devoted to explicating the underpinning theories of PBL paving way for an analysis of alignment between it and CAT teaching practice.

4. 2　Understanding theoretical foundations of PBL

In retrospect the PBL's development reveals to us that it was initially established on the basis of teaching practice before its theoretical roots received serious and systematic consideration (Graaff & Kolmos, 2003). Yet recent years have witnessed many efforts to situate this approach in social constructivism and cognitivism (Russell, Creedy & Davis, 1994; Savery & Duffy, 1995; Hendry, Frommer & Walker, 1999; Savin-Baden & Howell, 2004). The following section will disclose the foundation of PBL in social constructivism and cognitivism with special reference to the theories of ZPD proposed by Vygotsky (1978), and Bloom's pyramid taxonomy of learning objectives (Anderson & Krathwohl, 2001).

4. 2. 1   PBL with social constructivist view of knowledge and learning

Social constructivism is a theory pertaining to a wide range of disciplines, from philosophy, psychology to sociology and education, with its theorization generally attributable to Jean Piaget initially and later John Dewey and Lev Vygotsky (Oxford, 1997). In the educational circle, social constructivism is widely referred to as a theory of learning which has, though not always successfully, been translated into instructional design paradigms at different levels. Regardless of the divided findings as regards their effectiveness in teaching practice, the core assumptions underlying social constructivist teaching have been unanimously agreed upon which are found to be in remarkable consistence with those of PBL (Savery & Duffy, 1995).

First, social constructivist view of knowledge. Knowledge is believed to be constructed by individuals *in* their interactions with the environment. The implications include: firstly knowledge is not separable from the mind and therefore cannot be transferred from one person to another; rather students have to construct it by themselves. Secondly it is impossible to separate the knowledge from the context it is acquired. Knowledge is constructed by individuals when they interact and adapt to their environments. Thirdly knowledge is not an entity that is formed once and for all; instead, it is in constant change which is why some researchers argue for replacing *knowledge* with *knowing*, to give prominence to the dynamic nature of it, forming a postmodern concept of knowledge (Barrett, 2005). This view of knowledge is well manifested in PBL: first being a student-centred teaching approach PBL places responsibilities for learning on learners and makes teachers retreat to the background becoming facilitators and tutors providing support or guidance only as necessary. Second in light of the view that knowledge is inseparable from the environment where it is acquired, problems used to start PBL process are supposed to be authentic ones from real life under the assumption that knowledge acquired in solving the problems will be more readily transferred when under similar circumstances in their future work. Third, aware of the dynamic nature of knowledge, PBL is intended to develop in students the capability to extend and improve their knowledge against new challenges, holding cultivating lifelong learners as its primary goal.

Next, social constructivist view of learning. Learning is believed to take place when learners are engaged in social activities. Social negotiation and evaluation of the viability of individual understandings are critical for meaningful learning. By implication, first, learning takes place not only individually but also socially. Collaboration and group work are as a result essential in our learning and understanding as meanings are constructed in negotiating "within communicating groups" (Kim, 2001). We negotiate in our social environment with alternative understandings and additional information by which to construct knowledge based on viability. The other implication that can be drawn from here is that cognitive conflicts or puzzlement not only stimulate but also organize learning. As von Glaserfeld (1995) noted, other individuals are a mechanism to test our own views and the conflict arising out of the test would serve to stimulate new learning, hence enriching and expanding our knowledge. Again this view of learning is fully embodied in PBL tutorials: first, one of the basic elements of PBL tutorials is group work. Students are supposed to learn most effectively in groups of five to eight. Second, the tutorial has to start with a problem without any prior teaching in PBL, serving as the cognitive conflict or puzzlement that is believed by social constructivists to make learning necessary and possible. Third, students are given ample opportunities to negotiate their learning with group members, report to the class their learning outcomes and evaluate their newly acquired knowledge during each tutorial unit. Subsequently students' critical and reflective thinking skills are expected to be enhanced.

In sum, the analysis above reveals to us that PBL is an instructional approach in perfect alignment with social constructivism with built-in features embodying its view of knowledge and learning.

4. 2. 2 *PBL as a socially situated cognitive development process*

With a cognitivist perspective, PBL assumes that learning is a socially mediated cognitive development process.

First, according to PBL the students' prior knowledge is recognized as the starting point for instruction and learning (Bridges & Hallinger, 1995). The knowledge gap students find in tackling the problems between their prior knowledge and what they need to know that drives them to search for new information and necessary support. Additionally social mediation is critical in one's knowledge construction. Underlying PBL is the notion that learners' interaction with others, especially more knowledgeable or capable learners in society, contributes greatly to students' achieving new learning (Campbell, 1999). Learning is along this line considered a spirally cyclic cognitive progress forward mediated in society.

This view of learning links it up seamlessly with the constructivist theories of cognitive psychology, namely *ZPD* proposed by Vygotsky (1978), and *pyramid Taxonomy of Learning Objectives* (Anderson *et al.*, 2001). This idea of learning as progressive social-mediated cognitive development has been found to be adumbrated by Vygotsky's view in his well-known theory of ZPD (Chen *et al.*, 2009; Harland, 2003; Hoffman & Ritchie, 1997).

According to Vygotsky (1978), ZPD (see Figure 4. 3):

> …is the distance between the actual development level as determined by independent problem solving and the level of potential development as determined through problem solving under adult guidance or in collaboration with more capable peers. (p. 86)

ZPD, therefore, provides a theoretical framework for PBL learning progress. Fosnot (1996) points out that the ZPD "varies from individual to individual and reflects the ability of the learner to understand the logic of the scientific theory" (p. 19). This provides a theoretical basis for the design of PBL that is able to incorporate the

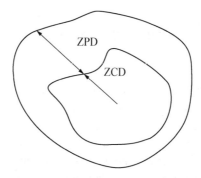

Figure 4. 3   Zone of Promimal Development（Vygotsky, 1978, p. 86）

Note: The Zone of Current Development (ZCD) represents the level that a learner can reach through independent problem solving and the ZPD as the potential distance the learner could reach with the help of a more capable peer. After successful instruction, the outer edge of the ZPD then defines the limits of the new ZCD.

different demands for learning of individuals with different levels of prior knowledge and for the choice of assessment methods that take into account also the individualized learning outcomes and the developmental perspective of knowledge evolution.

Besides, PBL parts itself from traditional transmissionist teaching approach not only in the notion about how learning takes place but also in the learning outcomes anticipated. Recent decades witnesses that higher education is shifting its demands from students' knowing *what* to knowing *how*. " Education is about equipping people with the cognitive and socio-emotional skills to be highly adaptable in fast-changing environments" (Tan, 2007, p. 102). It is against this background that PBL as an innovative teaching strategy has gained increasing popularity in a widening array of disciplines. PBL claims, with accumulating evidence, to be more capable of developing in learners higher order cognitive competencies (e.g. application, analysis, synthesis and evaluation) than the traditional pedagogy (Tan, 2007; Hakkarainen, 2009).

Therefore, Bloom's taxonomy and its revised version (Anderson & Krathwohl, 2001) have been widely adopted as a hierarchy framework by PBL designers and instructors to guide in their instructional design or a measurement tool of learners' learning outcomes. The revised version of the taxonomy is shown in Figure 4. 4.

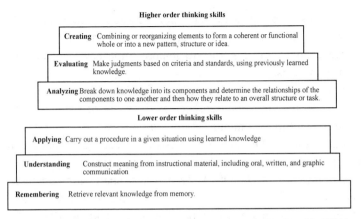

Higher order thinking skills

Creating Combining or reorganizing elements to form a coherent or functional whole or into a new pattern, structure or idea.

Evaluating Make judgments based on criteria and standards, using previously learned knowledge.

Analyzing Break down knowledge into its components and determine the relationships of the components to one another and then how they relate to an overall structure or task.

Lower order thinking skills

Applying Carry out a procedure in a given situation using learned knowledge

Understanding Construct meaning from instructional material, including oral, written, and graphic communication

Remembering Retrieve relevant knowledge from memory.

Figure 4. 4   Cognitive levels (revised by Anderson & Krathwohl, 2001)

4. 3   Connecting PBL with CAT teaching in higher education

In this section, an in-depth analysis will be presented of the state-of-the-art of CAT teaching in higher education in examination of the theoretical alignment between PBL and CAT teaching.

4. 3. 1   The new social demands for the CAT course

The pedagogical gap disclosed in Chapter 2 cannot be reasonably filled without a thorough knowledge of the demands of translator education in this new time and those of CAT teaching in the academic context. This section will therefore provide a closer look at those new demands which can in turn be translated into the educational goals of CAT courses in the context of higher education.

4. 3. 1. 1   Translator education at tertiary level

Biggs and Tang (2007) stressed the importance of a consistent message to be maintained across courses in a programme for the overall education quality. Being an integral part of translator education, CAT teaching should, therefore, be able to incorporate the demands for translators in general, which will then be examined first.

Looking back at the social changes in the past decade, those that have most bearing with translator education are unarguably globalization and digitization of translators' working environment. These two changes join together to impose some new demands for translator educators.

Among them, the first demand is that due to the explosive growth of information to be translated as the result of the globalized trade and culture communication, graduates of translation programmes should be able to translate increasingly specialized content in shorter timescales while maintaining high quality of their production (Alcina *et al.*, 2007). As a result, apart from linguistic competence traditional translation programmes have almost exclusively focused on a wider scope of specialized knowledge and a great variety of computer-based tools and resources have become must-have equipment for professional translators to meet this demand.

The second one is that translation is considered by more and more researchers a type of communicative competence which involves the ability of translators to interpret, express and negotiate meaning according to not only the task specifications but also social conventions (Colina, 2009, pp. 24-42). Translation students in this vein need to be trained in *self-concept* (Kiraly, 1995, p. 113) which is a mental construct including awareness of their social role in the profession, their professional duties and responsibilities, and knowledge of what others expect of them in their work. Colina (2009) therefore suggested that "the social aspects of translation have a prominent role in any methodology of translation teaching", "if translation teaching has as its goal the integral education of translators as professionals who not only perform and understand translation tasks but also function within a professional group" (p. 38).

Related to the above point is the third demand derived from the fact that the traditional mode of translators working individually in isolation has been largely replaced by teamwork in streamlined translation projects. Collaborative work involving close interaction with colleagues and clients enabled by computer-based tools is now an essential part of the translation process. Many researchers thus have advocated inclusion as a part of translators' competency profile of such competences as teamwork and accepting criticism as well as a well-developed expertise in technologies (Austermühl, 2001; Orbán, 2008, cited from Bilić, 2013).

The fourth demand is rooted in the modern translation profession which features rapid development and constant changes in both translation content and working environment. Thus translators' willingness to rise up to challenges and their ability to deal with new and complex problems are becoming increasingly important now when many researchers tend to adopt the notion that translation is in itself an activity of problem solving (Kaiser-Cooke, 1994; Wilss, 1996).

These new social demands for graduates of translation programmes are clearly reflected in the introduction into studies on translation teaching of the distinction between *training vs. education,* a pair of concept first distinguished by Widdowson (1984) in language pedagogy. By *training* he refers to the teaching " directed at preparing people to cope with problems anticipated in advance and amenable to solution by the application of formulae", while he believes *education* to be " developing general intellectual capacity, cognitive set, attitudes, dispositions which... can be subsequently brought to bear to deal with any eventuality that may arise" (p. 223). About the difference between *training* and *education* Moore (1998) also stated:

> *Training* to me means a narrowly focused program that leads to high proficiency in a specific skill. It prepares a student for one particular job or activity but provides neither broad perspective nor flexibility of approach. On the other hand, *education* enables students to see the forest *and* the trees. It encourages general approaches to problem solving and inculcates ways of thinking that are productive, effective, and rewarding. (p. 135).

This distinction is borrowed by Bernardini (2004) into discussion about translation pedagogy as to what should be taught of translation in different teaching contexts (long-term vs. short-term) and at different academic levels (undergraduate vs. postgraduate). According to Bernardini (ibid.), translation as an academic programme, at especially undergraduate level, should be aimed to *educate* prospect translators, that is, to develop their ability to "employ available knowledge to solve new problems and to gain new knowledge as the need arises" (p. 20). This opinion is echoed by more and more researchers. For example, Mossop (2003) stressed the university-based translation schools' function of fostering in students the general abilities required of by any slots that may arise in the language industry now and then, and Pym (2012) proposed lately the translation skill-sets that concern skills ("knowing how") rather than knowledge ("knowing what"). More recently, Bilić (2013) pointed out that translation programmes must adapt to the new reality of modern translation practice which requires not only a high language proficiency of translators but also expertise in TTs as well as social competencies such as teamwork and problem-solving skills (Austermühl, 2001; Orbán, 2008, cited from Bilić, 2013).

Similar opinions also began to appear among Chinese researchers of translator education though not explicitly conceptualized in this way. For example, Wang C. Y. (2012) observed from his survey findings that the three qualities translator employers value most are respectively *skillful mastery of translation tools and technology*, *continuous learning ability*, and *credibility and trustworthiness*. Besides, the objectives of undergraduate translation degree programmes (Zhong, 2011) included among them not only students' linguistic, cultural and technological competences, but also *critical thinking, communication and cooperation skills*.

To sum up, modern translation programmes in higher education have to address at least the following new societal demands for:

(1) mastery of computer-based tools and resources to ensure efficiency and quality of translation and facilitate teamwork in the translation process;

(2) awareness of and competence in translation as a social communicative task;

(3) willingness and ability to work in groups as well as individually;

(4) positive attitude towards challenges and adequate problem solving skills;

(5) self-motivated continuous or life-time learners.

4. 3. 1. 2   CAT teaching in translator education

The following section will probe into CAT teaching for its features in itself and the challenges it is faced with as they will determine the educational goals of a CAT course at the tertiary level.

As mentioned previously, publications about CAT training in the context of higher education are not many around the world with the situation even worse in China. Small in number though, researchers in the realm of translation teaching studies have begun to provide specific suggestions as to how to conduct CAT training as an integral part of translation pedagogy, especially in recent years when undergraduate translation programmes are mushrooming around the world.

Alcina is a researcher who has long been concerned with TTs and their teaching in higher education. The objectives of CAT teaching she suggests include making students "aware of the range of technology available to them today", "equip(ping) them with the knowledge and skills required to use these computer-based tools and resources as a valuable support for the tasks involved in the translation process" and "enabling future professional translators to judge which technological aid suits each situation best" (Alcina, 2008, p. 98). We can see in this set of the objectives not only declarative knowledge but also procedural skills and reflective thinking abilities.

In response to the aim of translation programmes "to produce translators ... with intellectual, professional and *technical skills*", Davies (2010) proposed *instrumentalisation* as one of the objectives of translator education and two standalone subjects to achieve it, i.e. subjects related to *Resourcing Skills* and *Computer Skills*. Aimed at a global conception, Davies, based on the findings in neurological research, suggested students at undergraduate level "acquire general skills with emphasis on procedural knowledge (knowing *how*)". As to the teaching method he favoured "meaningful learning and learner

111

autonomy" at all levels.

Samson (2010) is the only study devoted to CAT training at undergraduate level so far as the author could find. Seeing the enormous importance of computer skills in translation practice presently, Samson called for the inclusion of training of both general and specific technologies into the heart courses of undergraduate programmes. From the perspective of a teacher of four-year undergraduate translator-training programmes, he proposed a collaborative and student-centred teaching methodology. He argued that "with the presently distant and expanding horizons of knowledge available to students" it is no longer possible for teachers to assume a role of expert in all aspects of their field and thus the only authority as is in the traditional translation class. Instead, "collaborative work between a teacher and students … can restore much-needed excitement to the educational task" (p. 109). This methodology is also believed to be more workable when used with students with different starting points in computer skills as is always the case in reality. Besides, students in this way will enjoy more freedom to assess their knowledge and invest their time accordingly to suit their own needs which in turn would enhance their motivation.

What is most impressive is that Samson described in great detail a sample project of subtitling he used in his teaching and provides an inventory of projects (for training of different tools/resources) that may be adopted for CAT courses. The teaching effect was reported to be satisfactory as follows:

The dynamics of working in this way, with multiple digital information resources, flexible permanently open channels of dialogue, a project focus, together with class meetings for group instruction, guided study and tutorials, can help students move rapidly from a novice level, through introductory sessions and the establishment of a collaborative working environment, to more advanced autonomous modes of work that reflect professional practice." (ibid., p. 120)

Pym (2012) elucidated the translator skill-sets under the influence of the introduction of statistical MT into TM suite in translators' work. The analysis led him even as far as to shift the translator's role to mere *linguistic post-editing* which may not require "extensive area knowledge and possibly lay a reduced emphasis on foreign language expertise" (and thus his subversive comments on multicomponential models of translation competence). Although the revolutionary redefinition of translators' function is still subject to further deliberation, relevant to this current study is his call for attention to the fast rate of change in the field of TTs where all knowledge is provisional. Consequently he stresses that "knowing how to find knowledge" becomes more important than internalizing the knowledge itself and suggests more importance be attached to skills ("knowing how") rather than knowledge ("knowing that") in technology training.

In the above few yet intriguing discussions about CAT teaching in higher education, we can observe the following challenges to the CAT course and the needs they bring to the course:

(1) The constant update of CAT technologies and the explosive expansion of knowledge to be translated can never be exhausted by a single course. As a result, any attempt to teach specific operation and factual knowledge of a given tool may find itself in rapid obsolescence. As a result, this course is supposed to equip its students with not only factual knowledge of as many CAT tools and resources as needed but also the skills enough for application of them in real-life translation projects which involves the ability to analyse, evaluate and make judgment. This objective has gone far beyond the traditional translation course that is normally focused on linguistic competence;

(2) Students' ability to work and communicate with others has never been so important due to the modern work environment of translators on the one hand and, on the other, the intrinsic feature of this course that one of the major functions of CAT technology is to facilitate group work and distant communication. Thus team work spirit and ability ranks very high in this course;

(3) The equal accessibility of knowledge to students

threatens the teacher's status in the traditional teaching environment as the sole authority to impart knowledge and skills to students. Teachers in this course, therefore, should reconsider their role and the teaching method with an aim to guarantee their teaching effect.

4. 3. 1. 3　A Summary

As depicted in Chapter Two, CAT teaching has been widely acknowledged to be an integral part of translator education. Therefore, teaching CAT is not only a goal in itself but also in teaching translation. Consequently CAT teaching should serve, in addition to its own teaching objectives, the goals of the translator education as well. The author therefore probed for the new demands for translator education in general and challenges for CAT teaching in particular, the combination of which is expected to inform the pedagogical design of CAT teaching.

A comparative look at the results in the above search reveals a great deal of alignment in the challenges the new times posed on CAT teaching and translator education at college level that wait to be addressed urgently. They are as follows:

(1) Besides the necessary knowledge (declarative) and skills (procedural) of CAT tools and resources, the CAT teaching has to cultivate students' independent thinking so as to enable them to analyse, evaluate and judge the use and value of the technological instruments in actual tasks that are different one from another;

(2) Students in CAT courses should be trained in working collaboratively as well as individually as teamwork has been a dominant working mode of modern professional translators;

(3) Social aspects of professional translation have to be brought into the teaching to make students aware of their social roles and responsibilities as well as the professional norms which they will need to accomplish their future work as professionals;

(4) The constant change of CAT technologies and the intrinsic nature of instrumental competence as largely procedural knowledge combine to make it not only desirable but also necessary for this course to prioritize *how* over *what*. Students' problem-solving skills and self-motivation for continuous learning are therefore very important for them to keep up with the changes and become sensible users in real-life tasks;

(5) The CAT teaching has to be flexible enough to accommodate the varied needs from students who may have different starting points in computer skills, different initiatives, and different education backgrounds;

(6) The equal accessibility of knowledge to both students and teachers now has threatened the status of teachers as the sole authority to transmit knowledge to students as passive receivers. A new relationship between them along with a shift in teaching paradigm has been long called for in higher education at large.

4.3.2   Alignment between PBL and CAT teaching

A review of PBL core features and the new social demands for CAT teaching shows us a high rate of alignment between them. It allows us then to anticipate at least the following merits of adopting PBL in CAT teaching.

Firstly, from the students' perspective, PBL as a learner-centred approach to learning will give students much more freedom which may in turn help enhance learner autonomy and stimulate learning motivation and is more conducive to developing students' problem-solving skills as well as communication abilities that have been found necessary for future translators as reviewed above.

Students in PBL courses are not only allowed to decide what they are interested to learn, but also to determine how they want to study them. They are involved in assessing their own work and their classmates' work (Gallagher, 1997; Reynolds, 1997). In this way,

"〔S〕tudents develop a deeper awareness and ownership of important concepts in the course by working on activities, a basic tenet of the constructive approach to learning" (Seltzer *et al.*, 1996, p. 86).

The CAT course in higher education is responsible for turning out prospect translators who will have to be able to deal skillfully and flexibly enough with challenges on their future jobs with CAT technologies. This goal will not be achieved without the ability to learn independently and continuously (Davies, 2010). Worth more attention is that as CAT courses are more often offered in TTPs which in China recruit mostly art students, most of them are more or less resistant, if not hostile, to learning technology and things. Low motivation is always the most annoying issue teachers of this course have to deal with in order for students to obtain satisfactory results. Thus PBL's power to develop self-directed and self-motivated learners (Barrows & Kelson, 1995) and to stimulate students for learning and retain their interest hereafter will contribute greatly to the success of the CAT course.

Moreover, PBL courses normally start with authentic and ill-structured problems. Students are not expected to find the only correct answer to them but rather are encouraged to work out their own solutions based on the synthesis of their prior knowledge and new research findings with facilitation from tutors and other resources available. Students will benefit greatly from the PBL learning process which is focused upon problem-solving, as it helps develop both procedural and declarative knowledge and makes the student a skillful problem-solver. This makes PBL all the more suitable to a course of technology where students are not only supposed to acquire content knowledge but also operational skills. And the student's learning outcome as an efficient problem solver is exactly what is desired by both CAT educators and employers (Davies, 2010; Pym, 2012; Samson, 2010).

Furthermore, PBL will not only help students develop the subject-specific competence (i.e. CAT content knowledge as well as problem-solving skills), but also foster their general competences such as reasoning, communication, and self-assessment skills thanks to its organization of learning all in group work. This feature of PBL makes it particularly suitable to the CAT course seeing that most of the specialized CAT tools (a major part of teaching in this course) are designed in the first place to assist with translators' collaborative work because nowadays translation activities are more often than not administered in the form of projects (Gouadec, 2007; Kiraly, 2000) involving people of different functions and even situated in different quarters of the world. Even for freelance translators, communication and cooperation skills are their valuable adds-on for professional success. PBL is therefore expected to improve students' performance because the learning environment it promises is similar to the way students will be working later in the professional world.

Secondly, from the teacher's perspective, PBL can help free them from the unnecessary burden of constant designing and redesigning of their course totally afresh and from the dilemma about their role in class.

One of the greatest challenges the CAT teachers are faced with is the dynamics of the professional world. That is, as reviewed above, the TTs are in constant change which makes any knowledge about them provisional (Pym, 2012). Therefore, in the traditional objectivist teaching paradigm, to make their teaching aligned with the market needs, teachers have no choice but update or even renew altogether their teaching plan painfully frequently. Now this difficulty can be better addressed by the underpinning philosophy of PBL that takes an epistemological position seeing knowledge not as something static but rather something that is made and remade through *dialogue* (Shor & Freire, 1987, p. 100). Thus PBL is aimed to foster students' ability to *construct* knowledge in collaboration with tutors and peers in response to the emerging practical problems. This ability will equip them better for the constantly changing demands from work. In this way, the flexibility of PBL will free teachers from the impossible mission of the too frequent change in teaching just to keep up with the irrational pace of commercial products with an additional benefit to keep their teaching more coherent in both content and method.

Another trouble CAT teachers are confronted with is the dilemma about their role in the classroom. The author agrees with

Samson (2010) who argued that with the explosion of information on the one hand, and the more and more convenient access to knowledge available to students on the other, it is no longer possible for teachers to assume a role of expert in all aspects of their field, in this case TTs, and thus the only authority as is in the traditional translation class. Samson therefore proposed "collaborative work between a teacher and students" and claimed that such a relationship "can restore much-needed excitement to the educational task" (p. 109). PBL as a learner-centred learning approach saves CAT teachers from this dilemma by shifting teachers' role to facilitators and in this way reduces the teacher's power and increase learners' responsibility. Besides, this rebalance between the teacher and the student's roles is believed to create in students a sense of community which helps improve their communication skills and enhance their motivation.

Thirdly, from the perspective of the course as such, the author envisions the applicability of PBL to the CAT course on the account of its inherent nature as first an integral part of higher education and then a course to cultivate instrumental competence.

To begin with, PBL will benefit the CAT course by helping it educate lifelong learners. We are now living in what is called "society of knowledge" which in turn makes a "society of learning". Education at all levels has undergone reformation with an aim to equip students with an ability to do lifelong learning. "Learn to learn" has now taken the place of "what to learn" as a central concern of educators especially in adult education. The CAT course should strive for the same educational goal since it is an integral part of college education. PBL, with the assumption that learning is an active, integrated, and constructive process (Barrows, 1996), is targeted at developing students' intrinsic interest in the subject matter with emphasis on learning as opposed to recall in order to make students self-directed learners. A great amount of empirical evidence has shown that PBL is very effective for arousing and retaining learners' interest.

In addition, this learning approach lends itself to the CAT course perfectly also because, given the aim to develop students' *instrumental competence*, the CAT course would not be possible without a problem to start with since it is normally with a difficult problem to tackle that one needs tools. Furthermore, with each translation task being an ill-structured problem in itself, the tutorial process of PBL is in perfect alignment with the translation process and therefore shouldnot pose too much difficulty in the design.

Moreover, the rapidly increasing importance of CAT training for future translation professionals has compelled some researchers to call for a central status of it in translator education curriculum. It is not even uncommon to see translators referred to as *multilingual communicators* (Olvera-Lobo, 2007) or translation engineer (Gouadec, 2007). A pedagogical change in this course may have far-reaching influence on the overall learning outcomes. Further, the subject-level has been recommended (Masek, 2010) for initial PBL implementation as the feasibility and effectiveness can be better explored than in the broader programme context. The author, therefore, argues that the CAT course makes an ideal test field for initial attempts at a PBL approach to translator education.

To sum up, against the background of the changed societal demands for university graduates in general and the instrumental competence of professional translators in particular, the author has envisioned considerable advantages of using a PBL approach to CAT teaching at the tertiary level based on an analysis of the congruence between PBL and the attributes of the CAT course from the perspective of the student, the teacher and the course itself. PBL is, therefore, a justified choice of a teaching approach to the CAT course in higher education.

4. 4 A summary

This chapter has clarified the core features of PBL as an innovative pedagogical method by comparing and synthesizing a representative collection of previous studies. With a clearer understanding of PBL then, the theoretical connections between PBL and CAT teaching in higher education were established which paves the way to a pilot design and implementation later.

Chapter Five
The Designing Process

With a focus on the second stage of the study, this chapter will provide a detailed representation of the preliminary design of the PBL approach to CAT teaching in adherence with the concept and the principles of PBL as summarised in Chapter Four. Specifically, the pedagogical model of PBL revised from Barret (2005) (see Section 3. 1. 2. 2. 1) will be first operationalized by way of integrating a well-received instructional design model in constructivist paradigm, namely Constructivist Alignment (Biggs & Tang, 2007) and Campbell's (1999) PBL-focused instructional design model that has been preliminarily theoretically validated. The latter two models make the design framework applicable with not only the well-specified design elements but also the structure and the order of the designing. Then following it, the designing of each element will be reported in turn.

5. 1 The instructional design framework for a PBL approach to CAT teaching

As mentioned in the previous chapter, no PBL approach is readily available for CAT teaching in higher education. An instructional design framework using PBL at course level is therefore desired before systematic designing of a CAT course can really take place.

5. 1. 1 The framework building

As specified in Chapter Four, PBL is defined in the current study as a pedagogical strategy with two defining features, three major components and eight ground rules, which, when implemented right, are expected to achieve five educational goals (see Section 4.1.2). The author revised the PBL framework proposed by Barret (2005) with a hope to provide a general characterization of PBL as is visualized in Figure 4.1. Yet without exact structure and order to follow in the designing, the framework is not feasible enough for the instructional design. To fill in the gap, Constructivist Aligning (CA), a model of instructional design in constructivist paradigm, comes right in handy, with its consistency with the underlying philosophy of PBL and high workability, and Campbell's (1999) design model for PBL (see Figure 5.1) with its focus on specifics in PBL instruction design and preliminary theoretical validation.

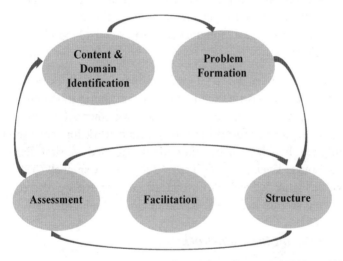

Figure 5.1   Constructivist PBL Design Model by Campbell (1999, p. 123)

CA is a framework for teaching design based on "the twin principles of constructivism in learning and alignment in the design of teaching and assessment" (Biggs & Tang, 2007, p. 52). The idea of keeping aligned between what is expected to turn out of learners, what is present in the teaching activities, and what is to be assessed has actually long before been found to enhance greatly the learning effects (Cohen, 1987). It is brought back to life by CA by joining it with the newly flourishing constructivist learning theory.

In addition to the designing principles to abide by, CA also specifies the exact procedures to follow in the designing process as demarcated in Biggs and Tang (2007, pp. 54-55):

(1) Describe the ILO in the form of a verb (learning activity), its object (the content) and specify the context and a standard the students are to attain;
(2) Create a learning environment using teaching/learning activities that address that verb and therefore are likely to bring about the intended outcome;
(3) Use assessment tasks that also contain that verb, thus enabling you to judge with the help of rubrics if and how well students' performances meet the criteria;
(4) Transform these judgments into standard grading criteria.

Campbell's model (1999) was proposed to operationalize particularly the PBL instruction design and contributes to the current model building with the more specified design elements for PBL.

Lastly, the author formed the design framework for the current study inspired by the combination of the revised version Barret (2005) at course level, Biggs & Tang (2007) and Campbell (1999), as shown in Figure 5. 2 below. Further explanation will be offered in the following section before the actual design is displayed afterward.

5. 1. 2   *Notes on the framework*

5. 1. 2. 1   A general introduction

As Figure 5. 2 shows, the PBL design process consists of *PBL course design*, *PBL tutorial design* and *PBL assessment design* which are all supposed to be well aligned within themselves as well as with each other, with the whole process governed by a common set of *social constructivist* and *cognitivist* theories. Besides, the three stages form a cycle starting but not ending with *PBL Course Design* which means that the instructional design works in a cyclical iterative

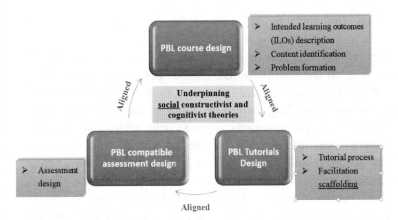

Figure 5. 2   The Instructional Design Framework of PBL Approach to CAT Teaching〔Based on Barret（2005）, Biggs & Tang（2007）and Campbell（1999）〕

fashion. The design could be strengthened and improved when new experience or knowledge gathers (Campbell, 1999).

At the outset of the cycle is the stage of PBL Course Design in which the researcher includes Learning Outcomes Description (Barret, 2005; Biggs & Tang, 2007), Content Identification and Problem Formation (Campbell, 1999). After it, when the goals are set and teaching materials ready, the tutorial pursues where tutors and students interact to make effective learning happen. In this case, a constructivist learning environment has to be created in which learners do self-directed study individually as well as collaboratively with tutor facilitation scaffold properly. So at the stage of PBL Tutorial Design, how the Tutorial Process is organized and how Facilitation Scaffold is provided should be specified. Then the last stage is PBL Compatible Assessment Design where assessment methods have to be worked out in alignment with the ILOs so that they will help engage students more in appropriate learning activities and optimize their learning outcomes.

5. 1. 2. 2   Stage 1: PBL course design

5. 1. 2. 2. 1   ILOs description

Among the three elements at this stage, describing the ILOs is made the first thing to do as it, on the one hand, dictates what content to be selected for teaching and, on the other, informs the learners of what and how they are to learn in the course. Here worth noticing is the fact that the author adopted Biggs and Tang's (2007) *Learning Outcomes* to replace *Teaching Objectives* which instruction designers have been long used to. By doing so the author wanted to reveal not only the underlying constructivist philosophy but also indicate a totally different way of phrasing when putting them into words. As Biggs and Tang (2007, p. 70) emphasized, the learning outcomes have to be phrased from students' perspectives, that is, with statements of what and how students are to learn instead of what topics teachers are to cover. One more thing worth noting here is that although *Intended Learning Outcomes* is used here, as Biggs and Tang (2007) argued, the teaching and assessment should "always allow for desirable but unintended outcomes" (p. 54) since in constructivist teaching learners are free to construct their own knowledge which may very well get beyond what is expected by the teacher.

In terms of the phrasing of ILOs, two suggestions deserve our attention. One is from Newman (2005) who provided a set of specific capabilities to be developed by PBL on the basis of the combined findings of Eagle (1991) and Woods (1995) (see Table 5.1).

Table 5.1   "Capabilities" that PBL Develops (Cited from Newman, 2005, p. 13)

· Awareness (active listening)	· Personal learning preference	· Defining real problems (goals, mission, vision)
· Problem solving	· Learning skills (laws, theories, concepts, etc.)	· Look back and extending experience (recognizng fundamentals in a given situation)
· Strategy (planning)	· Creativity	· Decision making
· Managing change	· Interpersonal skills	· Coping creatively with conflict
· Reasoning critically and creatively	· Adopting a more universal or holistic approach	· Practicing empathy, appreciating the other person's point of view

Continued

· Collaborating productively in groups or teams	· Selt-directed leaming	· Self-directed lifetime learning
· Self-assessment	· Obtaining criteria	

The other is from Biggs and Tang (2007) who suggested the SOLO (initials for Structure of Observed Learning Outcome) taxonomy, a hierarchy of understanding of knowledge deepening from *Unistructural* to *Extended abstract* stair cased by a series of verbs indicating different cognitive levels (see Table 5. 2). For more choices of helpful verbs, they refer researchers to Bloom's revised taxonomy (Anderson & Krathwohl, 2001) (see Table 5. 3). Provision of these specific verbs make the writing of ILOs much more manageable.

Table 5. 2   Some verbs for ILOs from the SOLO taxonomy (Biggs & Tang, 2007, p. 80)

Unistructural	memorize, identify, recognize, count, define, find, label, match, name, quote, recall, recite, order, tell, write, imitate
Multistructural	classify, describe, list, report, discuss, illustrate, select, narrate, compute, sequence, outline, separate
Relational	apply, integrate, analyse, explain, predict, conclude, summarize (précis), review, argue, transfer, make a plan, characterize, compare, contrast, differentiate, organize, debate, make a case, construct, review and rewrite, examine, translate, paraphrase, solve a problem
Extended abstract	theorize, hypothesize, generalize, reflect, generate, create, compare, invent, originate, prove from first principles, make an original case, solve from first principles

Table 5.3   Some more ILO verbs from Bloom's revised taxonomy

Remembering	define, describe, draw, find, identify, label, list, match, name, quote, recall, recite, tell, write
Understanding	classify, compare, exemplify, conclude, demonstrate, discuss, explain, identify, illustrate, interpret, paraphrase, predict, report
Applying	apply, change, choose, compute, dramatize, implement, interview, prepare, produce, role play, select, show, transfer, use
Analysing	analyse, characterize, classify, compare, contrast, debate, deconstruct, deduce, differentiate, discriminate, distinguish, examine, organize, outline, relate, research, separate, structure
Evaluating	appraise, argue, assess, choose, conclude, critique, decide, evaluate, judge, justify, predict, prioritize, prove, rank, rate, select, monitor
Creating	construct, design, develop, generate, hypothesise, invent, plan, produce, compose, create, make, perform, plan, produce

Source: Anderson and Krathwohl (2001)

5.1.2.2.2 Content identification

This step is intended to delineate the content of the course or, in Biggs and Tang's (2007) term, selecting the *topics* to cover in the course. When doing this, the designer needs to make highly demanding judgment on the breadth of the coverage and the level of understanding intended (Biggs & Tang, 2007). Many factors may affect the decision. For example, different teaching purposes of the educator or the grade the course is offered to. So fully considering all the affecting factors and striking a balance between breadth and depth are key to selection of appropriate content for achieving the intended teaching purposes.

5.1.2.2.3 Problem formation

As the principles of PBL design stipulate, the PBL course

126

centres around problems, though the terms used may vary from one study to another (e.g. problem, scenario, trigger) (Newman, 2005). To avoid confusion, *problem* is used in the current study. Problems in PBL are set to not only stimulate learning by activating learners' prior knowledge and arouse their intrinsic motivation to learn, but also provide a rich context in which learning takes place (Newman, 2005). Thus in the current PBL instruction design, problem formation is used as the counterpart of designing teaching/learning activities in CA (Biggs & Tang, 2007). The problems designed are supposed to incorporate fully and precisely the content to be taught in alignment with the ILOs.

Particularly for the design of problems in a PBL course, the summary of previous studies in the preceding chapter informs us of the following rules that have to be adhered to, in terms of respectively the problems' function, their nature, presentation and delivery.

First, the function of the problem. The problem in PBL serves to stimulate and support learning rather than simply elicit a solution. That is, the problem is not used to test skills and knowledge; it is used to assist with the development of skills and knowledge or it is where learning takes place (Boud & Feletti, 1991).

Second, the nature of the problem. The problem should be ill-structured and of real-life type. Being ill-structured means that the problem should be messy and complex in nature so that there is no single correct solution to it. Being real-life type means that the problem should be typical of the profession and is similar to those problems learners will likely encounter in future practice. Normally solving it requires integration of knowledge and skills across subjects as well as positive attitudes and motivation (Woods, 2000). In this way knowledge and skills acquired in solving the problem may then be readily applied to new problems that pop up in learners' future career.

Third, the presentation of the problem. Formats of problems may vary greatly depending on "the nature of the information to be acquired" (Charlin *et al.*, 1998, p. 4) and "the length of study" (Barrett, 2005, p. 17). Barrett (2005) provided some examples of different problem formats as shown in Table 5. 4.

Table 5. 4　Some different problem formats（Barrett，2005，p. 17）

Scenarios	Video clips	Physical Objects
Dialogues	Photographs	Letters
Cartoons	Poems	Metaphors
Diagrams	Limericks	Requests
Set of playing cards	Audio-tape recordings	Posters
Dilemmas	E-mails	Briefs
Progressive disclosure	Follow-ups	Quotations
Newspaper articles	T.V. shows	Literature

Fourth, the delivery of the problem. The problem should be delivered without any prior formal teaching. Learning in PBL should be stimulated and then organized and promoted by problems progressively in accordance with learning stages towards the goal of producing both the desired subject-specific knowledge and skills and the generic self-directed and motivated lifelong learning skills. To achieve this goal, problems should fall into students' capability to solve or be adapted to students' prior knowledge. They should also be of relevance and interest to students showing similarity to those they may encounter in their future professional practice. Besides, to engage students in discussion and assure expected learning outcomes, it is also critical that problems be delivered in forms suitable and understandable to students, making the context and the expected work clearly identifiable by students (Wee, 2004; Woods, 2000).

As problems play the foremost role in PBL teaching, problem design indisputably deserves the greatest attention. As Jonassen & Hung (2008) suggested, "utilizing appropriate type of problems to provide the students with appropriate contexts as well as the unique characteristics of that type of problem is critical for ensuring the effectiveness of PBL instruction, and in turn, optimizing PBL students' learning outcomes" (p. 17).

5. 1. 2. 3   Stage 2: PBL tutorial design

PBL tutorial design consists mainly of two elements, namely the tutorial process and facilitation scaffolding.

5. 1. 2. 3. 1   The tutorial process

Among all the elements in PBL instruction, the tutorial process is the least arguable one. Three characteristics have almost been kept common among different applications, namely cyclicality, self-direction of and collaboration in learning during the process (Hung, 2011).

First, about the cycle of learning in PBL tutorials. Researchers have reached an extensive agreement on what should be done during each cycle though the very steps specified by them vary from each other, among which are, for example, the four-step cycle reviewed by Hung *et al.* (2008), the six-step cycle defined by Barrett (2005), the seven-step cycle used in the McMaster-Maastricht PBL model (Graaff & Kolmos, 2003), the eight-step cycle summarized by Newman (2005), and the nine-step cycle built by Kirikova and Šveikauskas (2007).

Yet a comparison of these proposals of the learning cycle in PBL reveals to us a great deal of commonality. The two models from Newman (2005) (see Figure 5.3) and Kirikova & Šveikauskas (2007) (see Figure 5.4) are given below as an example.

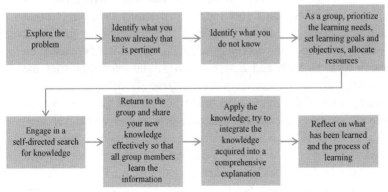

Figure 5.3   **The eight tasks of PBL learning cycle by Newman (2005, p. 15)**

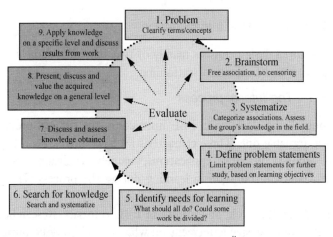

Figure 5.4　The PBL learning circle by Kirikova & Šveikauskas（2007, p. 3）

Regardless of the differing wordings in these two proposals, the only difference between them is the last step in Newman's version, that is, the reflection on the learning process and integration of the new learning into the learner's learning system.

Other than the steps to follow, the tutorial process has to guarantee student-directed learning, although in different models the degrees may vary. To achieve self-directedness of learning, tutors and learners would have to redefine their roles when transiting to PBL from the traditional teaching method. In PBL, learners do not receive knowledge passively in class from their teachers. Rather they are supposed to initiate and direct their own learning. Self-directedness could be realized by allowing students to set their own goals and objectives, decide their learning schedule, and identify based on research what to learn and how to learn it, and even how they are assessed. Tutors, inversely, would retreat from the forefront, providing help of different forms only when needed by the learner. It is agreed by most researchers that the tutor is less visible yet very critical in PBL. What is in traditional teaching paradigm referred to by *teacher* is replaced by *tutor* in PBL teaching given his/her changed role. The tutor's responsibilities are no more to provide lectures by which to transfer knowledge to students, telling them right from wrong in their thinking and *oughts* and *ought-nots* in their learning.

The tutor and the learner form a cognitive apprenticeship (Collins *et al.*, 1989, cited from Hmelo-Silver & Barrows, 2006), where learning is situated in complex problems facilitated by the tutor. In other words, tutors in PBL are expected to be a facilitator, resource guide and consultant to cater flexibly to arising situations, enabling students to carry out their research to collect and analyze information, make discoveries, and report their findings. (Aspy, Aspy & Quimby, 1993, cited from Tai & Yuan, 2007).

Through self-directed learning, students are expected to acquire not only lower-order ability to analyse and apply but also the higher-order ability to evaluate and reflect which will enable them to become life-long learners. Among the above-mentioned abilities, reflection has been emphasized by many researchers as being vital for PBL to achieve its educational goals.

What is also worth mentioning here is that the emphasis on self-directedness of the learning process does not preclude tutor guidance. Appropriate amount of scaffolding and learning resources provided by tutors in proper ways is necessary for turning out the expected outcomes (Hung, 2011; Kirschner, Sweller & Clark, 2006) which will be elaborated in the next section.

The third deciding characteristics of the PBL tutorial process is that learning is supposed to take place most effectively in groups. Small groups are therefore unanimously a must form of learning in any models applying PBL. Group work often undergoes the process as Feletti (1993, cited from Miller, Imrie & Cox, 1998) describes:

> As a group the students will identify topics or issues which later become the focus of their independent studies. The group may subdivide the list, work individually or in small teams, and later share their new information. As a part of the group process, students may try to explain new ideas or reason together how mechanisms work or systems interact. (p. 184)

People differ in their opinions about the ideal size of a group, but studies tend to show that "the development of skills for communication, the development of knowledge, and collaboration are best fostered in groups with between 5 and 10 members" (Newman, 2005, p. 16). Groups in each tutorial may meet with and without the tutor, and in some models with one member learning performing secretary and one chair. Cooperation, discussion, negotiation, and evaluation in group work are believed to be critical for meaningful learning to take place (Schmidt, Rotgans & Yew, 2011; Slavin, 1990; Springer, Stanne & Donovan, 1999).

5. 1. 2. 3. 2   Facilitation scaffolding

This section is aimed to explicate the design element of *facilitation scaffolding* in a PBL course. First in order is the explanation of the terms, their origins and denotation.

As mentioned above, PBL's emphasis on self-directedness in learning has led to considerable controversy over the necessity of teaching for this approach. It was misunderstood by many researchers (e.g. Kirschner *et al.*, 2006) to be "minimally guided" instruction where students have to discover and learn totally independently from classrooms or teachers. Yet arguments against such a faulty assumption have been published with evidence, making it clear that PBL engages students cognitively "in sense making, developing evidence-based explanations, and communicating their ideas" (Hmelo-Silver, Duncan & Chinn, 2007, p. 100) given the stress of constructivist learning theory on the importance of students being engaged in constructing their own knowledge. This engagement of students does not, however, exclude guidance altogether. Instead, as Hmelo-Silver *et al.* (2007) showed, PBL employs various forms of guidance, including even the old-fashioned *lectures*, yet different from the transmissionsist instruction, only on "a just-in-time basis" (p. 100). Taylor & Miflin (2008) also pointed out that self-direction for lifelong learning should be better conceived to be a goal "rather than the entire learning process by which it is achieved" (p. 753).

The author was convinced of the necessity for *guidance on a just-in-time basis* in PBL instruction as "it makes the learning more tractable for students by changing complex and difficult tasks in ways that make these tasks accessible, manageable, and within student's zone of proximal development" (Rogoff, 1990; Vygotsky,

1978, cited from Hmelo-Silver *et al.*, 2007, p. 100). Teacher guidance and self-directed learning are actually two sides of one coin, leading to a common goal — a life-long learner in a profession. Only, in this study, *facilitation* is the term chosen to designate the guidance offered by the tutor to be in consistency with the terminology used in PBL-related publications.

Scaffolding is an idea closely related to the concept of ZPD by Vygotsky (1978) and developed from the work of Jerome Bruner, who defines it as:

> a process of "setting up" the situation to make the child's entry easy and successful and then gradually pulling back and handing the role to the child as he becomes skilled enough to manage it. (Bruner, 1983, p. 60)

The notion then was developed in the pedagogical context to be a concept of both structure and process, referring not only to the relatively stable structured support but also the actual collaborative work that is being carried out. Walqui (2006) schematizes scaffolding in education as three related pedagogical "scales" as follows:

Scaffolding 1	Planned curriculum progression over time (e.g. a series of tasks over time, a project, a classroom ritual)
Scaffolding 2	The procedures used in a particular activity (an instantiation of Scaffolding 1)
Scaffolding 3	The collaborative process of interaction (the process of achieving Scaffolding 2)

Figure 5. 5   Schematic representation of scaffolding (Walqui, 2006, p. 164)

Besides this theoretical model for PBL, Barret (2005) provided pertinent guidance on its operation. For example, the operational definition of "PBL as a learning process" (p. 15) can be readily used to guide the tutorials; the PBL "curriculum matrix" (p. 17) is highly helpful in a PBL curriculum design, ensuring the alignment between the problems and the expected learning outcomes; the "checklist of problem formats" (p. 17) provide PBL designers a variety of selections for their problems and "the practical advice on ways to be an effective PBL tutor" (p. 19) gives teachers of PBL a benchmark to evaluate their performance against.

Along the scales, we can see the sequence moving "from macro to micro, from planned to improvised, from structure to process" (Walqui, 2006, p. 164). This scheme will be adopted as a framework in facilitation scaffolding in the current study abiding by the principle that teachers provide "contingent, collaborative and interactive" assistance (Wood, 1988, p. 96) for students in their solving problems which would be otherwise too difficult for them (Quintana *et al.*, 2004). While collaborative and interactive are quite self-obvious, being contingent here means that scaffolding depends on student actions, being flexible and reflexive as the whole teaching process in PBL is premised on the notion of tutors "handing over" the responsibilities to learners who take them over. As students' ability to solve problems develops, tutor facilitation will be adjusted or even dismantled. Anyway the purpose of this part of design is to scaffold facilitation "to help students engage in sense making, managing their investigations, problem-solving processes, and encouraging students to articulate their thinking and reflect on their learning" (Tiantong & Teemuangsai, 2013, p. 47) so that they may construct successfully the knowledge as expected.

5. 1. 2. 4 Stage 3：PBL assessment design

The last critical element in this framework is the assessment method, which is believed to be one of the most controversial issues in PBL (Hung, 2008; Savin-Baden, 2004). It has been noticed that great reliance on traditional exams of early implementations of PBL may have impaired the learning outcomes (Gijbels *et al.*, 2005) and begot the mixed and confusing research findings about the effectiveness of PBL. This finding led to wide recognition that assessment methods should be consistent with the educational goals of PBL and thus make for the desired learning outcomes of this instructional approach (Newman, 2005; Masek, 2010). Meanwhile recent years have fortunately witnessed a shift of emphasis in assessment from factual knowledge to application and transfer of knowledge.

A number of different assessment methodologies have emerged

to evaluate not only factual knowledge acquisition but also, more importantly, "students' problem-solving skills, reasoning skills, and personal progress" (Hung *et al.*, 2008). Various assessment formats have been devised, such as peer assessment (Papinczak, Young & Groves, 2007), self-assessment (Hung *et al.*, 2008), questionnaire survey (Valle, Petra, Martínez-González, Rojas-Ramirez, Morales-Lopez & Pinña-Garza, 1999), practical portfolios (Oberski, Matthews-Smith, Gray & Carter, 2004), Modified Essay Questions (MEQ) and the Triple Jump Exercise (Newman, 2005), and reflective journals (Ertmer *et al.*, 2009).

Among the searched studies on assessment design in constructivist paradigm, Biggs (2007) is the most systematic and highly employable one which will be then used for guidance in the current study.

According to Biggs (ibid.), an appropriate assessment task (AT) should be well-aligned with the ILO(s) the course is intended to address and be able to measure how well the learner has achieved it/ them. More specifically he provides five principles in designing assessment tasks:

(1) The criteria for the different grades, assigned to describe how well the assessment tasks have been performed, should be clearly outlined as rubrics that the students fully understand;

(2) One assessment task may address several ILOs;

(3) One ILO may be addressed by more than one assessment task;

(4) In selecting assessment tasks, the time spent by students performing them and by staff assessing students' performances, should reflect the relative importance of the ILOs;

(5) An important practical point is that the assessment tasks have to be *manageable*, both by students in terms of both time and resources in performing them and by staff in assessing students' performances. (ibid., pp. 195-196)

Besides, Biggs also summarised groups of typical declarative and functioning knowledge verbs by SOLO level that can be harnessed in assessment design (see Table 5.5).

Table 5.5 Some typical declarative and functioning knowledge verbs by SOLO level (Biggs, 2007)

	Declarative knowledge	Functioning knowledge
Unistructural	memorize, identify, recite	count, match, order
Multistructural	describe, classify	compute, illustrate
Relational	compare and contrast, explain, argue, analyse	apply, construct, translate, solve near problem, predict within same domain
Extended abstract	theorize, hypothesize, Reflect and improve, invent, create, solve unseen problems, predict to unknown domain	—

5.2　The preliminary design

5.2.1　The context description and model selection

As is widely observed, PBL implementations vary more or less in different educational settings to meet different instructional demands and contextual constraints (Taylor & Miflin, 2008). Many factors are believed to have caused these variations, such as the nature of disciplines, the learning goals, and the learner characteristics (Hung, 2011). Inevitably, these variations would generate different demands for students' cognitive and psychological engagement, which may in turn lead to varied learning outcomes. In view of the possible causes for confusion in understanding and evaluation of PBL, Hung (2011) drew on the previous attempts at categorization of PBL variations by Barrows (1986), Hmelo-Silver (2004), and Harden and Davis (1998), and identified six representative PBL models using Barrow's taxonomy as a structural framework as shown in Figure 5.6.

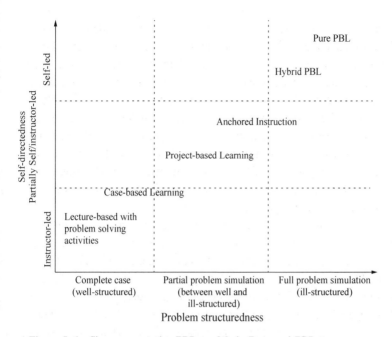

Figure 5. 6   Six representative PBL models in Barrows' PBL taxonomy
(**Hung, 2011**)

Thus to guarantee comparability of research on PBL, the detailed description of the context and the subsequent model chosen are advised to be included in PBL design (Hung, 2011). This is, therefore, the very first thing the author is going to do in her report of the preliminary design.

The context will be introduced mainly in terms of the course and its place in the curriculum of the programme, the facilities availability and, most important of all, the characteristics of the students. At the end, the level of application will be chosen in Huang's model (2011) based on the combination of the contextual constraints and the purpose of the current study.

1. The context of the CAT course under study

The course to be renovated with a PBL approach is computer-aided translation (CAT) in Department of Translation and Interpreting, an undergraduate translator training programme under English major at SIS of SYSU. The course of CAT is offered to students as an elective in the first term of their third year when students begin to take another course of translation, namely *Basic Translation (English to Chinese)*, concurrently. It is arranged in this way with the hope that by familiarizing them with the actual professional practice in use of specialized CAT tools right from the very start they learn translation, the students would accept the technology part of being a translator more readily, and consequently feel comfortable with and benefit greatly from the integration of CAT into their translation study. For the past 6 years, it has been offered in the traditional teacher-led lecture-based paradigm with an aim to introduce students to CAT concepts and practice.

To facilitate the course, the school set aside for it a laboratory equipped with 60 computers accessible to the Internet and installed them with two CAT systems — Oriental Yaxin and SDL Trados 2011. The former is a leading CAT system in China and the latter a top one internationally. Being a domestic product, Yaxin enjoys the advantage of flexible upgrading and customizing which is impossible for Trados. That is why the exact version of the Yaxin system is not and cannot be specified.

Speaking of learners characteristics, in the context of Chinese higher education, the following four characteristics are prominent and are believed to be common across different universities.

Firstly experience has it that students of this grade normally have little knowledge about CAT, not to mention experience in using it in practice.

The second one is that the students have been long taught in the transmissionist way since their primary and middle schools. Naturally majority of them are used to receiving knowledge passed down from teachers. Some of them are even resistant to attempts at turning them to be more autonomous learners (based on the author's personal experience in her attempts to introduce PBL ideas into her teaching previously).

Thirdly, born post 80s, most of them have begun to use ICTs very early in their life. Yet nearly six years of teaching CAT tells us that while largely competent in use of general electronic tools, they are still largely ignorant of any translation-specific tools and their functions in modern translation profession.

Lastly, with the majority of the students pursuing translation studies in China being females with arts education background, it is not uncommon to find, especially at the initial stage, somewhat resistance towards the learning of technology. Therefore, how to motivate students and incorporate different levels of students' computer skills is an important issue to consider for CAT teaching designers.

The context of the course, as described above, shows us some features that make for the adoption of PBL as a teaching approach (e.g. the availability of the necessary facilities and the low motivation of students to learn technology) on the one hand, and on the other points us to a potential hindrance, that is, the students' long experience and consequently heavy reliance on traditional transmissionist learning approach. Therefore, what awaited us ahead was presumably both promise and challenge, for which the author needed to be fully prepared.

2. Model selection

As the current study is an initial exploration of a PBL approach to CAT teaching, the author decided to push a little further so as to make the effects more observable and reveal more of the potential problems. Therefore, a *pure PBL model* was adopted which have the following features according to Hung (2011):

(1) In format, learning is initiated by a need to solve a real world, ill-structured problem, no lectures;

(2) In PBL process, problems solving reasoning is led by learners, content knowledge is self-acquired by learners, knowledge acquisition and application are carried out simultaneously, and the whole process is inquiry-based and highly contextualized;

(3) In problem design, the problems used in PBL are highly ill-structured;

(4) In learning outcomes, it is theoretically expected that learners will attain medium to high efficiency of content knowledge acquisition/coverage, very high ability to apply and transfer knowledge acquired, very

high level of problem-solving and reasoning skills, very high level of self-directed learning skill and very strong ability to cope with uncertainty.

5.2.2 The CAT course design

5.2.2.1 Content identification and ILOs description

This part follows Biggs and Tang's suggestion (2007, pp. 71-72), using the following two steps before the actual course ILOs are phrased:

(1) deciding what kind of knowledge is to be involved;
(2) selecting the topics to teach and decide the level of understanding desirable for students to achieve and how it is to be displayed.

5.2.2.1.1 The knowledge to be involved

As graduate attributes provide good guidelines for design of programme outcomes (ibid.), in the case of course outcomes, a recap of the social demands on translation trainees is necessary here.

As summarized in Section 4.3.1.3, CAT teaching in higher education is supposed to cultivate in students not only knowledge about *what* (e.g. what is CAT and what CAT technology is available for what use) but, more importantly, knowledge about *how* (e.g. students' ability to analyse, evaluate and judge the use and value of the technological instruments in actual tasks). Besides, at a more general level, learners' ability to collaborate and to pursue continuous study lifelong independently are also highly desired of translation graduates.

In sum, this course therefore involves both declarative and procedural, or, in Biggs & Tang's term, functioning knowledge. According to Biggs and Tang (2007), *declarative* or *propositional knowledge* refers to *knowing about* things or *knowing what* (p. 72). This kind of content knowledge is acquired through research from books and teachers' lectures, and can be tested by having students

" declare it back ". *Functioning knowledge* refers to that of performances based on understanding, or *knowing how*, and can be acquired but through experience. *Declarative knowledge* and *functioning knowledge* are believed to be separate from each other with the latter presupposing a solid foundation of the former.

5. 2. 2. 1. 2   The content selection

To decide on what to be involved in this course, the author relied on two major sources of information: One of them is from the professional point view, including an extensive review of literature review (as seen in Chapter Two) and a market survey she carried out searching for professional advice on CAT training in higher education (Luo, 2013). The overlapping part between them would be given primary importance while the new or stressed content peculiar to either of the two resources is also taken into account, but to different degrees. The other source is from the learner perspective. A brief survey was among the participating students to elicit the students' prior knowledge about CAT and what they wanted to learn from the course with the following three questions:

(1) What do you know about translation tools and their functions?
(2) What do you expect to learn from this course?
(3) Ask whatever questions you have about the course.

The responses from the students informed the author of the present knowledge level concerning CAT and provide, therefore, important clues for the author to decide the way to strike a balance between coverage and depth of the topics to cover.

The topics on the final listing are shown below in the order of their relative importance with types of knowledge to be acquired indicated in the braces:

(1) Narrow-sense CAT systems: mechanics and their application (declarative & functioning);
(2) Broad-sense CAT technology: use of corpus and search engines in translation (declarative & functioning);
(3) Technologically based translation project management

(declarative & functioning);

(4) MT: concept clarification, its modern development as well as its application in translation profession (declarative).

Then taking into consideration the contextual constraints, particularly the time limit and student background, the author managed to strike a balance between coverage and depth when selecting the specific content for each topic, as shown in Table 5. 6. In the table, we can see the four major topics to be covered in this course listed top-down in the order of importance as corresponding to the different lengths of classroom instruction. Narrow-sense CAT systems, taking nearly half of the overall instruction hours, clearly occupies the central position in the course. Its centrality is unarguably supported by the research findings as well as the survey results. SDL Trados 2011, Oriental Yaxin, and GTT are chosen based on a consideration of variety and availability. Firstly while SDL Trados 2011 and Oriental Yaxin are of desk-top and commercial type, GTT is web-based and freely accessible by all. In addition, SDL Trados 2011 is the well-recognized top brand among CAT systems worldwide and Oriental Yaxin is among the leading products in China. More than being of different origins, these two systems enjoy different target users, the former widely employed in international businesses and the latter by Chinese governments at different levels and state-owned enterprises. Therefore, in spite of the limited time in the course, the three systems are included for students to have a comprehensive view of the state-of-the-art of CAT technology; yet to compensate for depth, the students will be given freedom to choose to master only one of the three systems, leaving the other two optional depending on the learners' time and interest.

Technologically based translation project management ranks the second on the listing given the fact that translation practice is now largely carried out in the form of projects enabled by technology.

Table 5. 6   Content of the CAT course

Topics	Level of Importance	Types of Knowledge	Specific Content	Length of Classroom Instruction*
Narrow-sense CAT systems: mechanics and application	1	*declarative & functioning*	*SDL Trados* 2011, *Oriental Yaxin*, *Google Translator Toolkit* (GTT)	16
Technologically based translation project management	2	*declarative & functioning*	Simulate translation projects	10
Broad-sense CAT technology: use of corpus and search engines in translation	3	*declarative & functioning*	BYU online corpora, Google Search	6
MT: concept clarification, modern development and application in translation profession	4	*declarative*	Status-quo of MT technology with examples from online MT systems and application cases	2

﹡ The 2 more hours missing from here will be used for an orientation to familiarize students with PBL.

Then BYU online corpora and Google Search are included in broad-sense CAT technology for two reasons: For one thing, as students in translation programmes are largely language majors who are weak in subject matters which are critical for successful translation practice. This is the very place where broad-sense technology, particularly corpora and search engines, is highly helpful for. For the other, in the times of information explosion, new information and knowledge comes out every day. The ability to study continuously and independently is a must-have skill for anybody who expects sustainable development in his/her career. Among the very tools that could facilitate translators' lifelong learning are again corpora and search engines, with the former providing up-to-date language aid and the latter an inexhaustible source of general and expertise information.

Lastly, MT is included but expected to be understood only as declarative knowledge with a two-fold purpose. For one thing it is very often confused with CAT. So clarification of the two concepts is necessary for learners to understand CAT more precisely. For the other, MT, after going through a knockdown period between 1970s and 1980s, came back, with revenge. Thanks to the advances in computer science and different approaches to understanding languages (i.e. from rule-based to statistically based), MT technology has made great progress and been widely used for different purposes by individuals as well as corporations. Knowledge of its current development is therefore very important for learners to gain an unbiased view of TT. Nevertheless, without MT products nor tutors with relevant expertise, this part of teaching consists only of theoretical introduction.

5.2.2.1.3　The ILOs for the CAT course

After setting down the content of the teaching comes the most important step of the design — ILOs description. As the expected fruits of a PBL approach to CAT training in higher education, the ILOs should be phrased following the principles:

(1) The ILOs have to be compatible with the educational goals of PBL (see Section 4.1.2.2.4);
(2) The ILOs have to be phrased from the learner's perspective;
(3) The ILOs have to be phrased in such a way that "the student, when seeing a written ILO, would know what to do and how well to do it in order to meet the ILO" (Biggs & Tang, 2007, p. 71);
(4) The ILOs have to fully cover the selected teaching content.

Under the guidance of the principles above, using the verbs from Tables 5.2 and 5.3, the result ILOs of the CAT course with a PBL approach are as shown in Table 5.7 below.

Table 5. 7   The ILOs of the CAT course

After successfully completing the course, students will be able to:

1. Define CAT and MT, name a few leading systems of each and describe their mechanics, functions, development and applications against the background of the translation profession;
2. Explain the differences between CAT and MT;
3. Operate skillfully at least one narrow-sense CAT system among SDL Trados 2011, Oriental Yaxin and Google Translator Toolkit, to the effect that it can assist with his/her translation practice effectively;
4. Find and classify broad-sense CAT tools available to the translator;
5. Apply skillfully those broad-sense CAT tools covered in this course in translation learning and practice;
6. Explain what a translation project is and how it is normally managed;
7. Analyse a given translation task/problem with a view to selecting appropriate methods/tools for it;
8. Evaluate critically the usability and efficacy of adopted tools and resources and make well-grounded judgment about their values and suitability in different working contexts;
9. Plan a technologically assisted translation project as needed and reflect on its effects at its conclusion;
10. Communicate and cooperate with peers efficiently to achieve a common goal;
11. Monitor and assess the performance reasonably during solving a translation problem both individually and collectively;
12. Learn CAT proactively and continuously.

5. 2. 2. 2   Problem formation

Key to the success of PBL, the effectiveness of problems stays at the heart of any PBL implementation, atop all the issues to consider during the design process (Duch, 2001; Hung, 2006; Trafton & Midgett, 2001). In this section, the problems are designed and reported under the 3 strands of guidance, namely the design principles summarized in Section 5. 1. 2. 2. 3, the ILOs set down above and the student needs.

First, Section 5. 1. 2. 2. 3 has identified the underlying principles for problem design in terms of its function, nature, presentation, and delivery method. To recap, the problem in PBL should be an ill-structured and real-life one, delivered before any teaching takes place with progressively increased difficulty in accordance with learners' capability to not only stimulate but also organise learning which consists of both declarative and functioning knowledge acquisition.

Then, according to CA, the candidate PBL problems should be aligned with the ILOs in terms of not only breadth but also depth as, for one thing, the ILOs can help designers set the scope of the problem in light of the overall curricular standards and, for the other, the breadth and depth of the content afforded by the problems can be better balanced (Duch, 2001; Hung, 2006).

Lastly, besides the pre-course survey to elicit the students' self-aware knowledge and questions about the course (see Appendix 8), the author also collected via email from the students the most impressive translation difficulties they had encountered and how they managed to overcome them. From these real life problems, the author could not only extract material for the design of the translation scenarios (which will be used in Problem 2 below) but also infer the learners' present capability concerning use of tools in translation practice, which in turn provides good clues for the decision on the proper difficulty level of the designed translation problems.

To visualize the function of the guiding principles, a matrix of alignment was adopted for each problem design and integrated to provide an overview in Appendix 9. Two patterns can be seen in the matrix: first, declarative knowledge is arranged to be acquired prior to functioning knowledge as performance of the latter is dependent on the former (Biggs & Tang, 2007); second, one problem may address multiple ILOs and one ILO may be addressed in different problems which shows the recursive and open-ended nature of education.

Besides, the problems here will be reported in the following structure adapted from Yamane (2006):

(1) An introductory statement;
(2) ILOs addressed by the problem;
(3) The facilitation for the topic;
(4) The written assignment.

5. 2. 2. 2. 1 Problem 1: CAT and MT technology — what, how and why?

The first problem (See Table 5. 8) was designed simulating the situation where a layman is first curious about and comes to understand how TT has come along to its present form against the background of the development of translation from a personal activity to an international industry. It is expected that in representing the history of translation activity the students can not only get to know *what* CAT and MT technologies are but also, more importantly, *how* they work respectively and *why* they are needed as the way translation changes over time in response to the changed social demands. Story telling is adopted as the form of the problem as it is similar to the actual experience in real life in knowing new things, presupposing sufficient details which may motivate students to do extensive research while at the same time allowing for individual freedom and creativity that are highly desired features of ill-structured problems. Facilitation in various forms are available yet advised to be accessed only as needs arise.

Table 5. 8 The first problem

PROBLEM 1	
INTRODUCTION	CAT and MT technology: what, how and why
ILO(S) ADDRESSED	ILOs 1 & 2
FACILITATION	Readings on CAT and MT, websites with CAT and MT application cases, self-initiated QQ group chat, peer support
ASSIGNMENT	*Format:* story telling *Problem:* Recent years I have come across these words and pictures often when I read about the translation industry. What on earth do they mean? How do they have anything to do with the translation industry, its history and the role translation technology performs in it? *Directions:* Try to tell a story about the translation industry and translation technology, based on your collective prior knowledge and what you can search

Continued

PROBLEM 1	
ASSIGNMENT	for from the reading base, library, and online resources. Share your story with the class. Compare your story with the others' and find out what you can learn from them. Rewrite your story and try to fill it with as many details as you can. Words: Translators; Computers; Internet; Tools; Resources; Cooperation; Market needs; Multilingual; Localization; Globalization; Cloud Pictures: 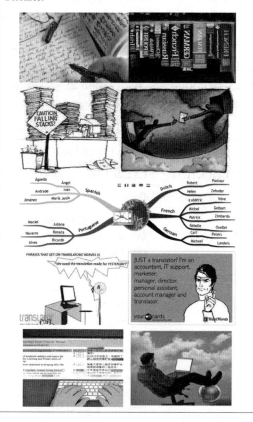

Continued

PROBLEM 1
ASSIGNMENT

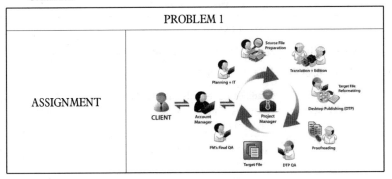

5.2.2.2.2   Problem 2: Broad-sense CAT-BYU corpora and Google Search

The second problem is intended to address the major broad-sense CAT tools. The literature review and student needs survey coincided in the most desired content to be covered concerning broad-sense CAT tools. That is, corpora and search engines. Among all that are accessible to general public, the BYU (Brigham Young University) corpora and Google Search are unarguably the best choices.

Since these tools are good assistance for translation, the best way of mastering them is using them in translation practice. Besides, the translation practice has to be neither too difficult nor too easy so that the students may find it, on the one hand, relevant to them and thus motivated to deal with it, and, on the other, cannot easily accomplish it without using tools. The author, therefore, collected from the participants via email translation problems they had personally encountered and found difficult to solve. Among the collected cases, the author then chose 10 that came up to the criterion and made 10 translation scenarios out of them. To cater to the possible different levels of capability among the students, the author added to the problem another 2 scenarios which are more demanding with longer texts and more "difficult points" (i.e. places in the text that can hardly be translated satisfactorily without using external help for the participating students).

Table 5.9　The second problem

PROBLEM 2	
INTRODUCTION	Broad-sense CAT-BYU corpora and Google Search
ILO(S) ADDRESSED	ILOs 4, 6, 7 & 9
FACILITATION	Readings on broad-sense CAT, websites with broad-sense CAT resources, self-initiated QQ group chat, peer support
ASSIGNMENT	*Format*: translation scenarios *Problems*: 1. 一个同学在翻译有关环境污染的新闻，其中出现"棕地治理"一词，根据上下文，了解到"棕地"是指化工厂搬迁后遗留下来的有污染物的土地，可是查阅了所有词典也没有找到这个词的翻译。他非常为难，你将如何解决这个词的翻译问题呢？ （Hint：这个问题可能有多个解决途径，你可以多尝试一些途径，并比较它们之间的效果差异） 2. 译者在翻译一篇小说时，曾遇到"green zebra"一词，从上下文可以猜出是种食物，但是网络上几乎没有相关的中文信息。于是译者询问了在美国的几个朋友，大概了解到了它是一种带有绿白色斑马纹的绿番茄，可是最后还是没有找到对应的中文官方名称，就直接译作了"绿番茄"。 请评价译者的这个翻译过程和译文选择。如果是你，你会怎么做？ 3. 一个译者在听政府报告时，发现翻译多把国债翻译成 treasury bond，很多场合下国债都被译作 treasury bond。有一次在字典中查找 treasury，发现多指政府发行的国库券，而国库券是指不超过一年的短期证券，所以他觉得将国债翻译成 treasury bond 是不准确的，应该使用 national debt。 你如何评价上述的思考过程？如果是你，你会怎么做？

Continued

PROBLEM 2	
ASSIGNMENT	(Limited in space, only 3 sample scenarios are displayed here. The full version of the 12 scenarios can be found in Appendix 10) *Directions*: Read 12 translation scenarios above carefully and try to solve the problems together with your group mates using the recommended tools and any other tools you find useful. You will have to share with the class your solutions and how the tools have helped you solve the problems later.

5. 2. 2. 2. 3   Problem 3: Narrow-sense CAT-SDL Trados 2011, Oriental Yaxin or Google Translator Toolkit

The third problem will enable students to acquire operational skills of at least one of the recommended three CAT systems by a simulated project (see Table 5. 10).

As explained above, the three systems are recommended to students with the consideration of variety and availability. Yet the students are free to choose any of them to master to the effect it can assist with their translation learning and practice. Therefore, in solving this problem, each group would have to make a comparison among the three products and decide for themselves with which one they would like to start.

The whole process of learning the chosen CAT system will be incorporated into a translation project which is exactly how CAT is applied in authentic professional environment and thus enables the most effective learning and skill transferability in their future work. Besides, to focus students' attention on learning CAT operation, the author takes two measures to reduce the workload of translation per se to the minimum: first the author chooses a brochure for a tourist site which is a genre quite familiar to most of the students and a text featuring simplicity and brevity. In addition, the tutor will provide students with a TM and a term base which have covered over 95%

Table 5. 10   The third problem

PROBLEM 3	
INTRODUCTION	Narrow-sense CAT-SDL Trados 2011, Oriental Yaxin or Google Translator Toolkit
ILO(S) ADDRESSED	ILOs 1, 2, 3, 9 & 11
FACILITATION	Readings on CAT, websites resources relating to narrow-sense CAT, video instruction of SDL Trados, self-initiated QQ group chat, peer support; TM and term base; external experts
ASSIGNMENT	*Format*: a simulated translation project *Problem*: Your group has just got a task to translate the brochure *Portugal's Undiscovered Blue Awaits You* into simplified Chinese. The brochure has 3, 839 words and you have to finish translating it within two days. The client offers you a TM and a term base. You are asked to do the translation on any of the CAT system — Oriental Yaxin, SDL Trados, or Google Translator Toolkit. *Directions:* Translate the brochure *Portugal's Undiscovered Blue Awaits You* into simplified Chinese (3, 839 words; See the sample pages in Appendix 11) in groups using any of the above CAT systems and employing an appropriate translation project plan. You'll have to go through the whole process of a translation project starting with a project plan and ending with a summative report on your gains and losses during the process. Cooperation among group members and use of technological aids (including broad-sense and narrow-sense CAT) are compulsory. At the end of the project, you will be asked to reflect upon the project and evaluate 1) your own and your group members' performance during the process and 2) the functionality of the CAT technologies you employed and suitability of them for the tasks addressed. You are encouraged to reach for more information and help via the above-mentioned means.

of the text to be translated. It incidentally serves another purpose of making students aware of the value of high quality TM and term bases for translation and as a result stimulating greater motivation for TM and term base building in their future learning process.

Anticipating great difficulty students might encounter at this stage, various forms of facilitation are prepared. Other than the readings and resources related to these three systems, human aids are strengthened. New to this stage, two external experts will be invited to join in the QQ group devoted to this course. They are both translation professionals who have rich translation and management experience. One of them is Mr. Wu, a translation manager in Grand Strong Limited (GSL), a Hong Kong-registered investment company, and the other, Mr. Xu, head of MT Division in SDL Shenzhen. Busy with their work, they cannot make it to the classroom in person, yet they both promise to provide in-time responses online every day during the whole course.

5. 2. 2. 2. 4   Problem 4: Translation project management and CAT

This problem takes the same format as the last one and will be used as both an assignment and an assessment by integrating all the required knowledge and skills addressed in ILOs in the course (see Table 5. 11). Also in the form of a translation project, here the author only highlights the differences between this problem and the third one.

With experience accumulated from the previous activities, students are expected to be more confident in using these tools to help with their translation practice. Therefore some changes to the facilitation are made. Firstly, this time the TM and term base are not directly provided by the tutor. Instead, the previous version of the user guide to be translated, namely *ApSIC Xbench 2. 9 User Guide EN*, and its translation will be given to students, out of which they will have to work out their own TM and term base for this translation task. Secondly, the tutor and the external experts are still accessible but are only allowed to provide guidance about how things can be done but not direct solutions.

Table 5. 11　The fourth problem

PROBLEM 4	
INTRODUCTION	Translation project management and CAT
ILO(S) ADDRESSED	ILOs 1-9
FACILITATION	Readings on CAT, websites resources relating to narrow-sense CAT, video instruction of SDL Trados, self-initiated QQ group chat, peer support; external experts
ASSIGNMENT	*Format*: a simulated translation project *Problem*: Your group has just got a task to translate the user guide *ApSIC Xbench 3. 0 User Guide EN* into simplified Chinese. The guide has 13, 033 words and you have to finish translating it within 10 days. The client offers you a TM and a term base. You are asked to do the translation on any of the CAT systems — Oriental Yaxin, SDL Trados, or Google Translator Toolkit. *Directions:* Translate the user guide *ApSIC Xbench 3. 0 User Guide EN* into simplified Chinese (13, 033 words; See the sample pages in Appendix 12) in groups using the CAT system you have mastered and employing an appropriate translation project plan. You'll have to go through the whole process of a translation project starting with a project plan and ending with a summative report on your gains and losses during the process. Cooperation among group members and use of technological aids (including broad-sense and narrow-sense CAT) are compulsory. At the end of the project, you will be asked to reflect upon the project and evaluate (1) your own and your group members' performance during the process and (2) the functionality of the CAT technologies you employed and suitability of them for the tasks addressed. You are encouraged to reach for more information and help via the above-mentioned means. (The full version of the problem can be found in Appendix 13.)

As to the text to be translated, the author chose one from an authentic task this time with the following reasons: First, the authenticity will motivate better the learners to do the translation as the author's personal experience tells. Second, a user guide is among the most suitable text types for a CAT project with 3 characteristics: (1) being a text about software, reasonably ample technical terms are expected in it which makes necessary a term base; (2) in such a technical text with a length of over 10 thousand words, certain amount of internal repetition at both sentential and terminological levels is possible which will show students the advantage of CAT in leveraging the former work by way of TM and term bases; (3) being a technical text, a software user guide is updated regularly with majority of its content remaining unchanged. This allows for highly efficient building up of TM and term bases out of the translated texts of older versions, which is also a very important function of CAT technology students need to master. Plus this length of a technical text with desirable amount of internal repetition also makes itself a best choice for group work, not only because every one of a group of 5 will get a moderate amount of work share, but also because the repeated content assigned to different translators will provide an opportunity for them to realize the advantage of CAT system in keeping inter-translator consistency.

Another difference is that the concept of translation project management will be stressed this time with specific requirements posed about the plan, organization and implementation of the project. Evaluation of the function of CAT technology in facilitating project management will be required as a major part of the summative report.

5.2.3   The tutorial process

5.2.3.1   The tutorial cycle

When deciding the exact tutorial process, the author was subject to two conditions: one is the ground rules set down earlier about PBL design (see Section 4.1.2.2 and Section 5.1.2.3) and the other is the actual context the design will be implemented in.

In terms of the tutorial process, it features cyclicality, self-direction of and collaboration in learning. More specifically there are 7 principles to abide by:

(1) Learning in PBL is problem-driven.
(2) Problems should be authentic and ill-structured as they are valued in the real world.
(3) Problems should be able to activate students' prior knowledge and provide meaningful context, relevance and motivation for learning to take place.
(4) Students are active participants who are responsible for and in control of their own learning.
(5) Students should work both independently and collaboratively in (small) groups in the learning process with facilitators providing necessary scaffolding and guidance.
(6) The learning process should be a self-regulated iterative process starting with problems, building on prior knowledge and skills.
(7) The learning process should include synthesis, integration of and reflection on learning and end with evaluation and review of the learner's experience.

Then as to the context of implementation, the top concern is about the student background. Especially in current situation when PBL is newly introduced into a course, as Hung (2011) pointed out, "explicit instruction to orient the student to PBL could help students start their PBL experiences with a positive attitude and the study habits that are aligned with the PBL process" (p. 545).

In accordance with the above principles, before the whole learning process kicks off, an orientation is in order for students to familiarize themselves with this new approach to learning. Then for each cycle of the tutorial process, the following steps are included:

(1) Students are presented the problem without any prior teaching.
(2) Students in groups analyse the problem.

(3) Students in groups identify their prior knowledge concerning the problem.

(4) Students in groups identify what they need to know about the problem.

(5) Students form their learning issues in groups and divide between them the possible work they will have to do toward a solution to the problem.

(6) Students engage in a self-directed search for knowledge, collecting and studying resources, and preparing reports to the group.

(7) Students share their new learning in groups and revisit the problem, forming new learning issues or rejecting some old ones based on their individual learning and discussion in groups.

(8) Students apply their newly acquired knowledge to the problem.

(9) Students review and reflect on their learning process and integrate their learning.

5. 2. 3. 2   The tutor role

As summarized in Section 5. 1. 2. 3, to ensure the self-directedness of the learning process, the tutor in this PBL CAT course will play a facilitator, resource guide and consultant to cater flexibly to arising situations, enabling students to carry out their research to collect and analyze information, make discoveries, and report their findings. (Aspy, Aspy & Quimby, 1993; cited from Tai & Yuan, 2007).

A further concern is about who is an ideal candidate for such a tutor role in PBL. It is not indisputable yet some basic criteria have been largely agreed upon. That is, the tutor is ideally expected to have both broad expertise in taught area(s) and rich experience in and skillful capability of learning facilitation. Broad expertise is necessary since PBL is intended to meet the needs of modern higher education for students with cross-disciplinary knowledge and skills (Barrow, 1996; Hung, 2011; Taylor & Miflin, 2008). Skillful facilitating capability is desirable because without it PBL risks running totally out of control or falling back to the old track dominated by teachers (Hung, 2011). How much guidance tutors should provide and the way guidance is given are still under debate. For example, in reflecting on how beliefs and values of teachers add confusion about what PBL is and can do, Taylor and Miflin (2008) revealed the different implementations all labeled with PBL where tutors are directive to greatly varied degrees and discussed different views as to whether lectures are a legitimate form of PBL tutoring. Therefore, instead of a definitive element, the exact tutor role is dependent on the actual context of a given PBL application as long as the baseline of being basically *supportive* is not tramped given that scaffolding or tutor guidance has been argued to be indispensable even in such a student-centred learning approach (Kirschner *et al.*, 2006; Taylor & Miflin, 2008).

The tutor in this course will abandon lectures altogether, showing up only in a responsive way.

5.2.3.3 The learner responsibilities

More specifically, in correspondence with the steps of the learning cycle, students need to identify initially their present knowledge levels and gaps in their knowledge and form for themselves learning issues as stimulated by the problem given. Then they need to solve the issues based (mainly) on their self-directed inquiry especially outside of the PBL tutorials. As Finucane, Johnson & Prideaux (1998) observes, the identification and pursuit of learning issues is a key feature of the PBL process. Last but not the least, the elaboration of knowledge via discussion and reflection on consolidated learning experiences at the last step of the cycle weigh greatly to guarantee learning outcomes and therefore was formulated as one of Schmidt (1983)'s three essential principles of PBL.

Yet what is worth mentioning here is that having been accustomed to the traditional teaching environment, most students would have to undergo a painstaking process of adaptation to the new learning responsibilities in PBL. Difficulties student normally encounter have been addressed by many researchers (e.g. Kolodner *et al.*, 2003; Moust, van Berkel & Schmidt, 2005; Perkins & Grotzer,

2000; Schimdt, 2000; Simons & Ertmer, 2005). These difficulties involve resetting both learning habits and mindset for being responsible for their own learning which are not reasonably expected to be accomplished in a short time and thus deserve tutors' persisting attention.

5. 2. 4   *The scaffolding*

This course adopted Walqui's (2006) scheme as a framework in facilitation scaffolding abiding by the principle that teachers provide " contingent, collaborative and interactive" assistance (Wood, 1988, p. 96) for students in their solving problems which would be otherwise too difficult for them (Duncan *et al.*, 2004). As Walqui (2006) classified, " in pedagogical contexts, scaffolding has come to refer to both aspects of the construction site: the supportive structure (which is relatively stable, though easy to assemble and reassemble) and the collaborative construction work that is carried out" (p. 164).

Along this line, this course employs also these two aspects of scaffolding: the structure one as seen in the format adopted for the problems, the sequence of the four problems and the learning process specified above, and the collaborative process one which is more improvised, initiated by students and responded by the tutor, providing learners with flexible help contingent on their needs.

More exactly, following van Lier's (2004) suggestion, four types of contexts were created for the process facilitation to take place in both " vertical construction and collective scaffolding" (Walqui, 2006, p. 167).

Thus, the student has available at least the following four sources of scaffolding (Walqui, 2006; see Figure 5. 7):

(1) being assisted by an expert, when the learner receives guidance, advice and modelling;

(2) collaborating with other learners, when learning is constructed together;

(3) assisting a lower-level learner, when both have opportunities to learn; and

（4）working alone, when internalised practices and strategies, inner speech, inner resources and experimentation are used. (p. 168)

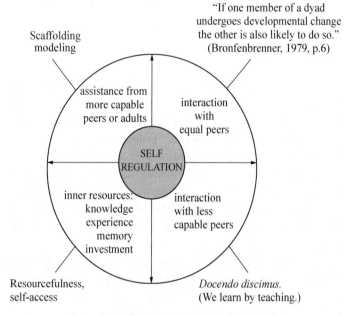

Figure 5. 7   Expanded ZDP（van Lier, 2004）

To provide a convenient platform for the vertical construction, some online resources can be of help, such as a QQ group and a Dropbox account. The former is a very popular chat room which is a good place for online direct communication and the latter a file hosting service whose cloud storage and file synchronization would make it quite easy to transmit or exchange files between either the tutor and students or between students themselves.

5. 2. 5   *The assessment design*

5. 2. 5. 1   Determining the guiding principles

Within the instructional design framework CA, the top guiding principle of the design of assessment tasks is the alignment between the assessments and the ILOs set down prior to teaching. As pointed out by some researchers, while teachers place primary importance on ILOs in an aligned teaching system, students usually allow themselves to be directed entirely by the assessments (Ramsden, 1992). This fact makes the alignment between assessments and ILOs all the more necessary as advocated by CA to produce more positive backwash (Biggs & Tang, 2007) as shown in Figure 5.8 below. By aligning assessments with ILOs, teachers and students can work through the learning process towards the same goals.

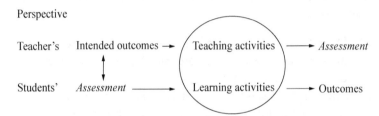

Figure 5.8   Teacher's and student's perspectives on assessment
(**Biggs & Tang, 2007, p. 169**)

Besides, using CA as the guiding framework of the design, the author is also supposed to consider carefully the following pairs of factors as specified by Biggs and Tang (2007, p. 191) before the final decisions are made in light of the ILOs of this PBL-guided CAT course. The result of the consideration is shown below (see Figure 5.9) that will serve a guideline for the choice of the proper assessment tasks and their grading methods.

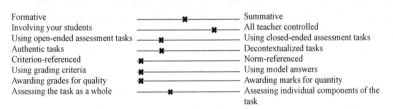

Figure 5.9   Where the assessments of this course stand
(**based on Biggs & Tang, 2007, p. 191**)

5. 2. 5. 2   Designing assessment tasks

In designing the assessment tasks, the ILOs, their different levels of importance and the manageability of the tasks for students are the three key factors for consideration.

With the ILOs intended to address not only declarative but functioning knowledge, this course adopted assessment methods that are process-oriented and authentic (Biggs & Tang, 2007; Tai & Yuan, 2007) which serve both formative and summative purposes. Besides, the suggestions from Biggs and Tang (2007) as to the forms of assessment are integrated and referred to when selecting the assessment tasks.

（1）Decontextualized assessments such as a written exam, or a term paper, which are suitable for assessing declarative knowledge;

（2）Performance assessments, such as a practicum, problem solving or diagnosing a case study, which are suitable for assessing functioning knowledge in its appropriate context; (p. 182)

（3）Assessments in higher education should include some open-ended tasks, namely divergent assessments, that allow for unintended outcomes, especially in PBL teaching with ILOs worded with "design", "reflect" etc. Tasks suitable for such a purpose are reflective journals, critical incidents and the portfolio. (p. 186)

A full view of the assessment tasks used in this course is displayed in Table 5. 12, which contains 6 formative ones and 3 summativeones. Below the author will explain briefly the ideas behind the adoption of such tasks.

First of all, the author took a process-oriented view of assessment or, in Biggs and Tang's (2007) term, "continuous assessment" (p. 164). That is, students' mastery of the required knowledge was not tested and decided till the end of the course by a single final exam. Instead, different measures were taken to keep a continuous record of student performance throughout the course,

Table 5.12   An overview of the assessment tasks in alignment with the ILOs

Assessment		Requirement		ILOs addressed
Formative	Problem 1	Understanding the key concepts concerning CAT and MT against the background of the translation industry	✓	ILOs 1 & 2
	Problem 2	Getting aware of and learning how to use skillfully broad-sense CAT tools and resources in translation practice	✓	ILOs 4, 5, 7, 8, 10, 11 & 12
	Problem 3	Getting aware of and learning how to use skillfully narrow-sense CAT tools and resources in translation practice	✓	ILOs 1-12
		Learning what translation project is, how it is conducted and why so	✓	
		Learning how to plan and implement a translation project facilitated by CAT technology	✓	
	Day-to-day performance in class meetings	Active participation with original contribution and logical & confident delivery, if needed in cooperation with other group members	✓	ILOs 1, 2, 4, 10 & 12
Formative & Summative	Individual study journal (daily)	What they study for this course	✓	ILOs 1, 2, 4, 10 & 12
		How they study	✓	
		Reflections on the effects of their study	✓	
	Group study record (per task)	What they study for this task	✓	ILOs 1, 2, 4, 10 & 12
		How they study	✓	
		Reflections on the effects of their study	✓	
Summative	Portfolio	Tools and resources they have collected during the course	✓	ILO 1, 4, 12
		Self-initiated CAT-related readings	✓	
		CAT translation practice	✓	
		Any other self-initiated efforts	✓	
	Problem 4	Covering all those under Problems 1-3	✓	ILOs 1-12
	Essay questions	Being familiar with the key concepts concerning CAT, MT and translation projects	✓	ILOs 1, 2, 4 & 6

which is expected to provide the teacher with a view of students' growth on an ongoing basis and help her keep the students on track toward the desired outcomes, and hopefully to produce better learning effects. Among them are the students' day to day performance in classroom meetings, and the documents including the individual study journal each student would have to keep every day, the group study journal that each group keeps for every task and *portfolios* each student keeps individually.

In both kinds of the study journals, students are required to note down at least what they study for this course or the task, how they study and reflect on the effects of their study (e.g. what they could have done better and how). In the portfolios, the students are required to note down every initiated efforts they make concerning CAT in learning or practice.

The other dimension of the assessments is their purposes. In the current study, Biggs and Tang's (2007) definitions of *formative* and *summative assessments* are adopted: Formative assessment is provided *during* learning, telling students how well they are doing and what might need improving; summative *after* learning, informing how well students have learned what they were supposed to have learned (p. 97). According to this definition, all the assessment tasks but Problem 4 and the essay questions to be answered at the end of the course are formative with the portfolio, the journals, and students' day-to-day performance also serving as summative tasks based on which the final mark is given. That is, most of the tasks serve a double-purpose, being at the same time a task in which the desired learning takes place and an assessment by which the tutor is informed of the learners' progress. As pointed out by Biggs and Tang (2007), such a double purpose of assessments may cause students confusion about whether to display errors to get rightful feedback from the tutor (for formative advantage) or to hide errors in order to get higher marks in the final grading (for summative advantage). To reduce the confusion to the minimum, two measures are taken. Firstly, the students are informed of the forms and purposes of all these assessment tasks at the very beginning of the course so that they could know when to do what. Then for the tasks serving the double purpose, a decision is made: it is not the documents and performance in themselves that are graded; instead it is the observable progress as evidenced in the documents and the daily performance that is graded. The grading of the progress will take up 10% of the final mark because it provides strong cases for the learner's enhanced motivation and independent learning ability (i.e. ILO 12). This way the author assumes that the learners can shift their attention from how many errors they make to how better they can do with the problems, which would drive the learner positively towards the desired learning then. The composition of the final mark is as Table 5. 13 shows.

Lastly worth mentioning is the adoption of the concept of authentic assessment or performance assessment (Biggs & Tang, 2007; Moss, 1992; Tai & Yuan, 2007), which is recommended by many researchers to be suitable particularly for PBL teaching with an intention to promote largely students' functioning knowledge of problem solving.

More exactly, by authentic assessment, the author refers to the assessment tasks that "represent the knowledge to be learned in a way that is authentic to real life" (Biggs & Tang, 2007, p. 180). That is why a translation project derived from a real translation task is used as the final assessment task.

5. 2. 5. 3 Forming the grading criteria of the summative assessments

In this part, the author will illustrate what are in those summative assessments measured and how. First of all, a concept has to be clarified before the exact grading criteria can be decided, namely the model of summative assessment used in this study.

As Biggs and Tang (2007) distinguished, measurement model and standard model of summative assessments are very often confused. According to them, measurement model is intended to compare and rank students normally quantitatively to cater to the purpose of selection while standard model is aimed to "assess the changes in performance as a result of learning, for the purpose of seeing what, and how well, something has been learned" (p. 170).

Table 5.13 The composition of the final mark

Assessment task			ILOs addressed	Percentage
Progress as evidenced in journals and day-to-day performance			ILOs 1, 2, 4, 10 & 12	10
Portfolio			ILO 12	5
Essay questions			ILOs 1, 2, 4, 6	10
Problem 4	Plan	Project plan	ILOs 6, 7, 9	15
		Technological aids	ILOs 1, 2, 4, 7, 9	
	Implementation	Process management	ILOs 1-5, 6, 10, 11	40
		Final delivery	ILOs 3, 5	
		Quality Assurance	ILOs 6, 9	
	Reflection	Technological aids evaluation	ILOs 7, 8, 9, 11, 12	20
		Reflective report		
		Experience exchange	ILOs 6, 7, 8, 9, 12	

(The 90 percentage total spans Portfolio through Reflection.)

The purpose of the latter is not to compare one student with another, but to find out how well what is taught has been learnt. Such an assessment, therefore, measures against criteria or is criterion-referenced (CRA) as opposed to the measurement model which is norm-referenced (NRA). As Biggs and Tang (2007) and Taylor (1994) argued, it is the standard model that fits into university education in which students are expected to enhance their academic competence and sharpen their performance.

This course will thus adopt the standard model and consequently adopt the following assumptions of this model:

First, teachers can set standards (criteria) as ILOs of our teaching. That means, when well-written, the ILOs can be used as criteria against which how well learners have met them can be assessed. With this model, it is the learner's performance not the person that is assessed and normally it is *how well* instead of *how much* the learner has learned. As a result, a hierarchy of levels is preferred to mere marks. Such grading can be very facilitated by the use of explicit rubrics (Biggs & Tang, 2007). The grading rubrics can be found in Appendix 14.

Second, different performances can reflect the same standards. This principle applies particularly to professional training where a problem can be solved in different ways. In such a course adopting a PBL approach in which every learning task, namely the problem, is open to different solutions, this standard model is therefore the best choice. The formats of the assessment tasks, such as a project and portfolios, are chosen here in line of this principle.

Third, teachers can judge performances against the criteria. This requires that the teacher should be clear about what constitute high quality performance and why before making sensible holistic judgment in the assessment. Again set-up of the grading rubrics will make this easier and more stable.

Thus different from in measurement model, assessments here are qualitative and assume different concepts of validity and reliability. In this study, validity resides in the alignment between the test and the *interpretation* and *uses* it is put to (Biggs & Tang, 2007). That is, if the test can enhance the claimed ILOs and the performance the

ILOs target, it is valid. Meanwhile *reliability* in this study constitute *intra-judgment* and *inter-judgment reliability*. The former means that the test is reliable if the same person grades the same performance in the same way on different occasions and the latter means that different raters grade in the same way the same performance on the same occasion (ibid.).

Then all the principles set, abiding by the standard model, guided by CA, students in this course will be assessed in terms of how well the ILOs are achieved with the rubrics of grading criteria built in advance as shown in Appendix 15. Evidence will be collected from the tasks in terms of the ILOs and judgment made holistically then. To compromise with the administrative requirement for marks, the summative assessments will be addressed as a whole and graded by levels which will then be converted into percentage points.

Two more points worth mentioning before the conclusion of this section are firstly the inclusion of self-assessment and peer-assessment in the summative assessment task, namely Problem 4, and secondly, the combination of assessment of individuals and groups in this course.

First, the final grade of this course relies heavily on the judgment about to what extent the criteria have been met. It has been widely argued that students can and should be involved in this process of judgment making starting from the setup of the criteria through selecting evidence the judgment can be made against to the final judgment making as to how well the ILOs have been met (Boud, 1995; Harris & Bell, 1986, cited from Biggs & Tang, 2007). Therefore, both self-and peer-assessment are employed in the current design upon the following two considerations:

For one thing, with the pressure to assess their peers as well as themselves, they will have a better understanding of the criteria and what evidence is good for meeting the criteria, which in turn engage them better in learning.

For the other, making judgment about whether a performance of one's own and others is good or not against given criteria is a vital ability in professional life. While the traditional education method disempower students in this regard, this course design using a PBL approach should and can play a better empowering role.

Second, the inclusion of assessing the group as a whole is almost inevitable since the PBL approach to this course stipulates that the whole learning process takes place. Group-based grading, therefore, makes for a context for students to achieve the PBL-desired ILOs. Such as communication skills, respect for others, collaboration skills. Besides, it has an extra advantage for the tutor in that the peer pressure can "reduce the likelihood of students failing to keep up with the workload" in his/her absence (Wood, 2003, p. 330).

5.3 A summary

In this chapter, the author demonstrates the instruction design she piloted using the PBL approach. The PBL model is first built and executed under the guidance of the instructional design theory, constructive alignment (CA). The three major components of the PBL approach are then materialized in a given context. In the next chapter, the initial effects of this design in preliminary teaching will be examined as perceived by the learners.

Chapter Six
Findings of the Preliminary Field Test

This chapter is aimed to report the findings from the preliminary field test of the designed PBL approach to CAT teaching as described in the previous chapter. The design was put to practice in SYSU during the first term of academic year 2013-2014 as a four-week credit-bearing project. As explained in Chapter Three, the effects of the teaching were preliminarily explored qualitatively via a triangulated method from the students' perspective since, being the initial stage of educational design research, the current study was aimed to understand, rather than judge, the process which validates the reliance on the insiders' experience. The following part will then present an overview of the findings first. Detailed description will ensue of the positive and the negative findings in turn concerning the teaching effects on the three levels of evaluation (Kirkpatrick, 1998).

6.1 An overview

6.1.1 A few notes to begin with

Before the final report is presented, a few notes related to the analysis are needed for a better understanding of the data analysis process. As Lapan *et al.* (2012, p. 255) advised, the author should start her analysis with revelation of the researcher position, researcher biases or predispositions which is believed to serve the following purposes: For one thing, the revelation makes it transparent to readers of this study the researcher's stand and thereby her limitations. For the other, more importantly, by this analysis, the researcher is herself highly aware of her own limits and therefore became more cautious when making any interpretations and conclusions.

In this particular case, the author had been teaching in college for over 10 years in the traditional transmissionist paradigm. During the years of offering the CAT course she had been deeply concerned about the efficacy of such a traditional teaching approach seeing that almost every time the course was over, all that the teacher heard was complaint about the course regardless of how hard the teacher had tried, to enrich the content or to diversify the activities. The complaint most often heard from learners was that they did not know after having learnt all this much why they had to learn it especially when they instantly forgot about what they had taken so much effort to learn. The PBL approach struck the author the first time she read about it with its overwhelmingly appealing power in fostering in learners' functioning as well as declarative knowledge and, more importantly, in enhancing their motivation for independent and continuous learning especially in professional education. The author, with her teaching experience of over 10 years, found herself identifying with the idea underpinning PBL that problems are the starting point of learning and it is in solving problems that learning takes place most effectively which can later be retained longer and transferred more readily in real life. This predisposition to embrace PBL into her teaching is what makes this study happen in the first place but meanwhile might also hamper her interpretation of the data. In view of this possibility for biased judgment, she would try her best to make the analysis process as transparent as possible in addition to such measures taken in research methodology as triangulation.

6.1.2 General description

As described in Sections 3.3.3.3.4 and 3.3.3.3.5 in Chapter Three, data collected were all fed into QSR NVivo 10 and analysed using mainly the theory-driven approach. Below (see Figure 6.1) is the screenshot of the input data organized according to source types, which can be further categorized into self-reported data, namely individual journals, group study records, individual reflective reports and the survey results, and factual data which include the summative assessment documents, tutor journals and classroom observations. When the data were ready, coding was conducted in the order specified in Table 3.4.

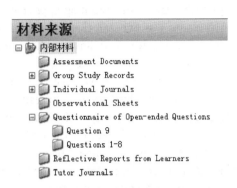

Figure 6.1　QSR NVivo 10 screenshot of the internals folder

The coded content were then analysed and compared between the two categories before the general findings were sorted out. The findings will be reported below with the 3-level evaluation framework (Kirkpatrick, 1998) before discussions are made based on the integrated view of the findings together.

6.2　Findings of the first level: reaction

In this section, the findings about the level of reaction will be reported mainly based on the self-reported data, namely the participants' reflective reports and their responses to the survey questions 1, 6, 7 and 8. Among the data coded with the first theme (see Table 3.4), we see both generally positive reaction and concerns as well as suggestions for improvement. Details will follow below.

6.2.1　Wholly positive reaction

Data of this level showed a wholly positive reaction to the PBL approach as all the 24 respondents of the survey expressed a high degree of satisfaction with the adoption of the PBL approach in the CAT course, with "much better than the traditional approach" appearing in 16 responses to the first question in the questionnaire and "better" in 8 of them. Very confirming to the author is that the reasons the respondents enlisted for their overall satisfaction with the PBL approach coincide highly with the expected outcomes of the instruction. More specifically the PBL approach produced among the learners "stronger motivation", "better ability to study independently", "better transferability of the learned knowledge and skills", "better efficiency in learning by group study", "longer retention of the learned knowledge", "deeper understanding", "more critical and logical thinking", "better communication skills", "more confident in her potential to learn technology".

Among the reasons the commonest ones for their favoring the PBL approach are the greatly enhanced motivation, the individualized and democratic learning process and the benefits of group study.

That the PBL approach enhanced greatly their motivation for learning CAT is mentioned in most of the participants' reflective reports. Two excerpts were quoted below:

[Quote 1][1]
"这样能激发学生的学习兴趣和主动性，充分体现出学习的【学】，应该是主动的。要求放在那里，是自己要去做到这些要求完成任务，遇到的问题也是自己的，要自己想办法解决，学习终究是自己的事。"
["*This (PBL) could stimulate students' interest and motivation. It makes us understand learners are at the centre of learning. Learning is not effective when we are only required to do it. It is only when learners take the challenge of the tasks proactively, meet and solve their own*

---

① Quotes from the data are marked in square brackets and sequenced numerically. Translation of the quotes was done by the author and attached below the original ones in braces. The translated texts are in italics to be differentiated from the body text of this thesis.

173

problems, that learning can effectively take place. Learning after all is the learner's own business."]

[Quote 2]
"更能够针对自己的问题学习，更有效果，主动性增强。"
["*(With the PBL approach), we could focus on our own problems. Learning in this way was more efficient. We had stronger motivation to learn."*]

As expressed in the second quote above, many participants also found this PBL approach more democratic in allowing learners more freedom and power to decide what and how to learn themselves through communication within groups and with the tutor. The more individualized learning process did not only benefit the learners in that it allowed them to focus on their own problems but also set the tutor free from the work that should not have been necessary. Two other quotes from the survey are given below:

[Quote 3]
"因为不同的人可以有不同的解决方法去完成一个项目，相互沟通和自主学习更【民主】发挥个性化；传统学习太僵硬，套在每个学生身上，都是被逼着被要求着按老师的方式学，以后碰到不同问题可能就不行了。"
["*Different students could solve the same problem with different methods. You could learn your own way and exchange your ideas with your group members. This way of learning is more democratic and could accommodate more individualized habits and features of different learners. The traditional approach of teaching, however, is too rigid. Every student has to learn in the same way as the teacher requires. Learning in that way can be hardly applied to life in the future when the problem they meet is different from that they have learnt."*]

[Quote 4]
"学生在自己探索过程中学习的知识更扎实且适用自己

的消化模式，老师省去低端重复讲解的步骤，可以留更
多时间讲解一些翻译发展的趋势、技术等专业高端的知
识。"

["*(With this approach), students explored their own
learning process. Learning as a result of independent
exploration was deeper and fit better with one's own
knowledge system (i.e. could be better assimilated into
one's existing knowledge system). The tutor could save a
lot of time from repeating those preliminary (unnecessary)
teaching, leaving enough time for more advanced topics
such as the trends in the translation industry and translation
technology.*"]

Among the self-reported comments on their overall feeling
about the PBL approach, what seemed to impress the participants
most is the group study. As shown in the following five quotes, most
of them expressed their appreciation of this learning approach:

[Quote 5]
"我感触很深的首先是这次的项目学习是以小组为单位
的模式，大家通过小组学习获取知识和一起完成项目，
这样的感觉和之前高中单打独斗个人奋战的感觉非常不
同。"

["*What impressed me most in this project is the group
study. The whole group learned and accomplished the
project together. It felt so different from how I learnt all
by myself back in middle school.*"]

[Quote 6]
"个人还是非常赞赏 PBL 的这种在小组进行的积极主动
的学习方式。印象很深的是，那天我自己上网查找一个
问题，第二天竟然还很清楚地记得那个问题的答案，或
许这就是填鸭式学习和主动学习的不同吧。"

["*Personally I like very much such an active and
explorative study in groups with the PBL approach. What
impressed me a lot was the day when I searched online for
solutions to a problem which I could remember clearly the*"]

next day. This, I guess, is the very difference between active learning and spoon-fed learning."]

［Quote 7］
"我一直讨厌技术较强的实践且消极地认为自己就是在这方面不行，但这次作业为了对组内负责，需要自己学习操作软件，我发现我虽然不是技术达人，但是在绝境下也是可以掌握技术并开发教学视频中没有讲述的功能的。让我看到了自己自主学习的潜力。"

［"*I had hated techniques and thought myself poor in this regard. But this time I had to learn on my own how to operate the software well in order to fulfil my share of responsibilities in our group. From this experience I found that without talent for technology, I could, when desperately, all the same manage to master the software and even explored some new functions of it uncovered in the teaching video. It convinced me of my potential for independent learning.*"]

［Quote 8］
"我比较适合规划，但是创新上还是不强。遇到问题很难转换思路解决，但是小组里的共同讨论弥补了这点。希望日后自己能有更多视角。"

［"*I am a person good at planning but weak at creating stuff. I am not flexible when dealing with problems and often fail to get them solved, which is no more a problem in this course, with group members to discuss together. Hope I can learn to look at things from different perspectives.*"]

［Quote 9］
"除了老师一定的指导，从周围优秀的同学身上，也学习到了很多，他们或是有独特的见解，或是优秀的台风，或是过硬的技术，或是整理资料的能力，或是犀利的问题。这些小小的细节和习惯都使我受益匪浅。"

［"*Besides some instruction from the tutor, I have learnt a lot from other classmates. Some of them had very insightful viewpoints, and some others showed great stage manners (when doing presentations). Some had great talent for technology and still some were very good at filing documents or posing incisive questions. All these had benefited me greatly."*]

6.2.2   Expressed concerns

Although admitting the overall positive effects of the PBL approach to CAT teaching in this project, some participants expressed concerns over the adoption of this new approach in the future. The mentioned concerns include:

(1) PBL is more time-consuming.

[Quote 10]
"自己探索印象深刻，但容易浪费时间。"
[" *(With the PBL approach,) we did explorative study and knowledge acquired this way can be remembered more deeply, but it tended to take too much time.* "]

(2) Possible "free riders" or "trouble makers" would hamper the effects of group study.

[Quote 11]
"但也会出现问题，尤其在小组合作中，如果有同学对某一环节的学习有所欠缺，则会导致整个项目出问题。"
[" *But (the PBL approach) also caused some problems, especially in group study. If some student cannot play his/her part well enough, he/she would put the whole group in trouble.* "]

(3) Too heavy after-class workload makes the PBL approach almost impossible to be implemented in the normal term with the same effects.

[Quote 12]
"不过课后的学习也会比较重一些，在正常学期进行的

话效果不知道能不能这么好。"

["*But the workload after class was heavier. So I doubt the PBL approach would produce equally good effects in the normal long term.*"]

[Quote 13]

"虽然我很喜欢 PBL 这种形式的学习，但它对个人学习的强度和精力要求也很高。如果放到平时，在要兼顾所有课程的前提下，实施起来比较困难。"

["*Although I liked the PBL approach to learning very much, it is highly demanding for the learner's devotion and energy. It may be much more difficult for it to be implemented in the normal long term when students have to deal with a lot of courses at the same time.*"]

(4) Both learners and teachers who have been used to the traditional teaching approach may find it difficult to adapt to the new approach and during the time it takes for adaption learners may suffer some losses and negative emotions.

[Quote 14]

"能够激发学生创造力，但也应考虑到学生在此转换过程中的不适应，如没有积极应用老师这一工具。"

["*(The PBL approach) could enhance students' creativity. Yet it deserves some attention that students may not adapt easily to such a new approach. Such slow adaption may cause them some losses (e.g. failure to make good use of the changed role of the tutor).*"]

[Quote 15]

"有趣的是，我们已经习惯去认同老师传递给我们的信息。印象最深刻的是第一节课的时候，老师放了 TED 视频后让我们提取关键词。我们组当时其实思路很乱，后来老师来了之后分析得很透彻，很'高深'，我们立即'恍然大悟'，立即改变了我们原来的思路方向，赶着奔着向老师给我们的思路前进。后来各组上台写自己的 KEYWORDS 时，我们发现我们组的最'有深度'，但又发现这其实不是我们自己的劳动成果，所以大家当时其实都很泄气。"

["*Interestingly we have long been used to accepting without questioning the information teachers pass to us. What impressed me most was the first lesson in which the tutor played us a TED video first and let us summarise what we had learnt from the video with keywords. Our group was actually in a mess when the tutor came to us with her own very in-depth analysis. Her idea was so 'insightful' that we were instantly 'inspired', and dropped what we were discussing and rushed towards the direction the tutor pointed for us. Later when the groups shared their KEYWORDS with each other, we found our results most 'profound', yet felt nonetheless discouraged realizing that those were not results of our own efforts at all.*"]

(5) What if learners lacked desired motivation?

[Quote 16]
"但有一个问题就是如果学生真的非常缺乏学习的主动性，靠小组的约束也没有，在小组合作中就会出现很多问题，而且他本人的收获也会比较少。我觉得如果这种教学方法要推广，就必须考虑起码是大多数学生的特点，这种也应该考虑在内。"

["*But there will be a problem if students lack real motivation to learn. Those without much motivation will cause the group some trouble regardless of the pressure from the within the group. They themselves would learn little. Therefore, the PBL approach will have to take into account such different student characteristics if it is to be successfully applied more widely.*"]

(6) The PBL approach to CAT teaching may be difficult to be implemented in big classes.

[Quote 17]
"但是或许这种方式在太大的班级里边还暂时难以实现，

因为这种模式需要学生比较大的热情。加上类似这种 CAT 课程的前期准备（比如 Trados 的安装、学习）会耗费比较多的时间、出现比较多的问题，如果学生太多，或许会出现难以应付的情况。"

［"*But maybe this approach can hardly be implemented in big classes, because learning in this approach requires great passion from the learners. Besides, in preparation for the course, we had to install Trados and learn how to operate it which all took a lot of time as many problems arose during the process. Too many students then would very possibly make this course impossible.*"］

（7）Less content knowledge may be acquired than in the traditional teaching paradigm.

［Quote 18］
"我最初的质疑就是以这种新的学习方法，强调通过问题引导和小组的自主学习，可能效率会比较低，学到的知识也没那么多。但在课程结束之后，说实话，我觉得可能学到的知识也许不如用传统方式来的多，但真的是一个宝贵的体验，有很多的经验教训，也表现出自己的不足。"

［"*About the new approach, I initially suspected that it stressed problem-driven, independent learning in groups and might not be efficient enough. Content knowledge acquired in this approach might not be so much as in the traditional way. Upon the completion of the course, frankly speaking, the result did prove me right. The knowledge acquired was less than in the traditional way, but such an experience was still quite precious to me, giving me a lot of lessons and revealing to me a lot of my weaknesses.*"］

What is interesting here is that all the concerns the participants expressed above have been discussed extensively in previous literature on PBL. Some of them have been found problematic and addressed in depth, such as the issue of implementing PBL in large classes by Shipman and Duch (2001) and suggestions as to how to facilitate learners' transition to PBL by Allen & White (2001).

6. 2. 3 Suggestions for improvement

In addition to the concerns, the participants also shared their opinions as to how to improve the design of the course in their reflective reports and responses to the questionnaire. The suggestions involve such different aspects of this course as scaffolding, the tutor role, grouping method, and assessment. Quotes are given below for each of this aspect.

(1) About scaffolding

Suggestions concerning scaffolding were the most, which reveals to the researcher, on the one hand, the participants' habitual reliance on the tutor for solutions as inherited from the traditional teaching approach and their failure to understand the principles of the PBL approach to education, and on the other, the need to further examine the ZPD of the learners and improve the way tutor guidance is provided among the four types of process scaffolding as designed in Section 5. 2. 4.

For example, in the following quote, the learner wanted the tutor to teach them how to solve problems before they set out to learn new stuff, which is exactly the way of learning in the traditional transmissionist paradigm. She did not actually understand the underlying learning theory of the PBL approach that there is not an only correct solution to a problem that can be learnt from the tutor and one cannot really learn how to solve problems until he/she solves problems him/herself. Besides, new knowledge was not supposed to be acquired by students who had got the "correct method" of problem solving from the tutor; rather, knowledge was believed to take place during learners' exploration for solutions to problems themselves, individually and collectively.

[Quote 19]
"一直以为这类学习方式需要适合这类学习方式的学生来配合。在学习前应该给一个指导，教会他们解决问题的方法，以免学生走许多弯路。"
[" *I had always thought that this kind of learning approach*

*fits best with students who are used to this particular way
of learning. But prior to turning to this way of learning
more guidance had better be provided to students teaching
them the methods for solving problems so as to save them
unnecessary trouble during the study."*]

The appearance of the above comment reveals to the author
that more explanation and guidance are needed concerning the new
learning approach at the beginning of the course, as suggested by
some other participants below.

[Quote 20]
"我们在前期基本处于迷茫状态，不知道这门课程要讲
什么，会学到什么。希望以后在第一节课时老师可以更
加明确地进行一些指导。"
[*" We were largely at loss during the first few days, not
knowing what this course was going to teach and what we
would expect to learn from it. We suggest that the tutor
give clearer guidance in the first lesson next time."*]

[Quote 21]
"翻译工具这门课难度大，入门门槛比较高，通过自学
的方法，一开始尝试安装软件还有看说明书，不知道是
讲什么，好几天都云里雾里很迷惑。而且技术不好，感
觉自己作为实验课的小白鼠，要被电脑折磨死了的感
觉。所以建议老师，能够先通过讲解入门，然后以后的
过程让大家自己探索和完成。PS：不过，通过讨论和询
问，拨开云雾见阳光的感觉，实在太棒了！"
[*" The CAT course is difficult with a high threshold. (It
was highly demanding) if we had to install the software and
read the user guide in preparation for its later use all by
ourselves. We had been very much puzzled during the first
few days, not knowing what it was all about. What's worse,
a very poor learner of technology, I felt like a guinea pig,
almost tortured to death by the computer. So my
suggestion to the tutor is that the course begin with some
elementary introduction by the tutor before the learners are*

*asked to explore and accomplish (the tasks) independently.
PS: However, it felt wonderful indeed to get everything
clear and done after all this discussion and enquiry
process.*"]

However, the tutor should be very cautious when providing such guidance because, as shown in the second quote above, sometimes learners' initial confusion may be actually a normal state at the early stage of problem solving instead of that about the learning approach. The confusion of the former kind is normally a must experience especially for green hands in professional problem solving and serves as a benchmark against which the learner may feel a great sense of achievement after overcoming all the difficulties and get things done, like the above student expressed in her P. S. .

Still some other learners wanted the tutor to include some traditional teaching approach in the course or mark a common deadline for everybody to finish certain and certain task. The following is an example of this kind:

[Quote 22]
"或许可以在某些环节采取传统教学的方式来讲述一些
要点。或在给的资源中标明在什么时间之前必须完成阅
读之类，可以保证大家的学习进度一致。"
["*Maybe somewhere during the course the tutor could
have explained some major points using the traditional way
of teaching. Or she could have marked deadlines for the
reading resources and demanded what reading materials be
finished before the deadlines so as to ensure a common
progress of the class.*"]

This quote suggests to the author, on the one hand, that this learner was possibly not completely adapted to the PBL approach yet. Therefore he/she still tended to rely on the tutor to tell him/her the important content, which would make him/her feel safe. Besides, also common in the traditional teaching approach, all the learners in a course would be asked to progress at the same pace, advancing too quickly or lagging too much behind would cause trouble for the tutor, knowing not how to accommodate different needs at the same time. Yet it is exactly where the PBL stands out since by allowing learners great freedom to learn in their groups, PBL does not only allow but encourage students to progress at their own pace as long as they could achieve the expected outcomes at the end of the course. Here judged from this suggestion, the student seemed to be disturbed when finding that the class were not progressing at the same speed, which again may be a sign of an inherited habit from the tradition teaching approach. Nevertheless, this kind of suggestion reminded the researcher of the necessity for more attention to the progress of the individuals especially during the early days, for one thing, to provide timely help for those progressing too slowly indeed, and for another, to keep oneself well-informed of the overall reaction of the class to the course design so that necessary changes or adjustment can be made in time to ensure its success.

As to the QQ group the author applied for the course serving as a platform on which the tutor provided the learners with process scaffolding (see Section 5.2.4), most of the participants found it useful, as the following comment shows:

[Quote 23]
"我觉得 QQ 群十分有帮助。1. 课程内同学的相互沟通，分享经验，或者预先告知之后会发生的问题以避免。2. 同学和骆老师的沟通，双向，及时。3. 和其他老师的沟通，是一笔资源。而且发现那些老师都好厉害啊!"
["*I found the QQ group quite helpful. First the classmates were able to communicate on it with each other, sharing experience and warning each other against some possible trouble so that it could be averted. Second with it we could have bilateral communication with Teacher Luo just in time. Third we could communicate with the invited experts on the group, which is a real legacy for us. Plus they are really something!*"]

But suggestions were also made in view of some shortcomings of it. For example, some student pointed out that the written communication on QQ was not as effective and efficient as verbal communication via telephone.

［Quote 24］
"有些问题以文字形式说明其实说不清，交流的效果有时候还稍有偏颇。"
［"*Some problems could hardly be explained clearly in writing and possibly caused misunderstanding.*"］

［Quote 25］
"缺点就是有时候同学们问的问题会淹没在消息内容中，还要自己去翻看聊天记录。最好是找到有那种消息置顶功能的通讯软件。"
［"*The shortcoming is that sometimes the questions asked earlier would be drowned in new messages coming up later. You've got to look through a bunch of messages just to look for the one you want. So it is better to find some communication application with the function of sticky-posts.*"］

［Quote 26］
"不好之处是问题与回答比较分散，如果同学有相同问题的话，可能会找不到聊天记录，造成重复问题。相比较之下感觉发帖子的形式更好一些。"
［"*What's bad is that questions and answers could be scattered (by irrelevant messages). It would be difficult to find whether a question has been asked and answered already and as a result the same question may be possibly asked again. Comparatively a posting application would have been better.*"］

Accordingly next time the choice of the platform for facilitation can be reconsidered taking these suggestions into account.

(2) About the way of grouping

The second commonest suggestion was about the way of grouping. More specifically, diversity in the features of members within groups and equality among different groups are two major factors to consider when grouping students as shown in the following quotes:

［Quote 27］
"各小组成员要有各自的特点，会有更好的效果。"
［*The learning effects would have been better if learners of more distinct characteristics had been grouped together.*"］

［Quote 28］
"各组人员的分配尽量能力平均。"
［*The average competence of each group should better be even.*"］

［Quote 29］
"还是应当根据学生的能力特征，较为平均地分配人。"
［*Students of different capabilities should better be evenly distributed across groups.*"］

(3) About the tutor's role

The next kind of suggestions involve the changed role of the tutor from the authoritative to the facilitator. The participants' suggestions below show us that we would better appropriate more time for their adaption to the new approach, making the shift of learning approach slower.

［Quote 30］
"老师的作用其实可以循序渐进地淡化，在初期应让学生对课程有清晰的认识，并让学生了解各种进度，在这些基础上，让学生进行自己的学习，渐渐进入 facilitator 的角色。"
［*The traditional role of the tutor would better be weakened step by step. It would be better if the tutor makes the students well informed of the course content and procedures in the beginning, leading them to learn more*

independently and retreats to be a facilitator gradually."]

[Quote 31]
"比如小组项目执行完全依靠组内成员的安排，老师指导较少。但这也有我们自己的问题，我们应该积极向老师寻求帮助。"
[" *For example, during the implementation phase of the project problem, we had to depend almost all on ourselves and got rather little guidance from the tutor. Of course, it also partly our own problem. We should have been able to get more help if we had approached the tutor more proactively."*]

Also what is worth attention here is that tutors should be alert when providing help, keeping in mind that the purpose of helping is to enable the leaners to study without the external help gradually. As shown in the second quote above, after such a short project, the learners were actually aware that the tutor should be taken as part of the learning resources to proactively reach for help instead of an expert to transmit knowledge to them. Thus tutors should keep an eye on the leaners' change in this regard and encourage them to engage more bravely in independent learning.

(4) About assessment
Suggestions were also made that more of the everyday tasks be graded and incorporated into assessment. An example is given below.

[Quote 32]
"规定小组之间的课下交流，不仅写个人的 journal，小组内的会议记录，将小组间交流的记录、感想也算入作业和评分，更大程度上加大同学们的互动沟通。"
[" *The tutor can demand that, to keep track of the communication within groups, learners keep individual journals, meeting minutes, records of group communication and their reflections. And these records should all be assignments to be graded. This way learners may feel more motivated to communicate with each other."*]

Such a kind of suggestion indicates to the researcher that some learners have been aware of the importance of motivation for learning effects and they do value very much the backwash effects of assessment on enhancing motivation.

(5) About the problem design

In the self-reported data, different problems were commented with or without sufficient supporting reasoning.

First, 3 out of the 24 participants found the first problem (Problem 1: CAT and MT technology — what, how and why?) confusing.

[Quote 33]
"其实不太清楚这一活动的意图，或者下次可以对您想通过这个活动来了解我们什么方面稍作介绍。"
["*We did not actually understand what this problem was for. Maybe next time you can explain more clearly to us what you want us to learn from this problem.*"]

The fact was that the tutor had explained in detail the reason for starting the course with this problem was for them to have a whole picture of where CAT technology is against the backdrop of the development of the translation industry. The keywords and pictures given to them marked the watersheds in translation history which were unexceptionally driven by the advancement of TT. Although small in number the suggestions also reminded the researcher of the need to slow down in the first a few sessions of the course when learners are struggling with the transition to a new learning approach which may distract their attention and results in poor understanding of the intention of the problems.

Then only 2 of the 24 participants found the second problem (Problem 2: Broad-sense CAT-BYU corpora and Google Search) not that necessary, thinking broad-sense CAT technology has been quite familiar to them. But since there are still 22 learners who did not show any objection to this problem, the author would take it as well-accepted.

Lastly, although few suggestions were made about the third and fourth problems, namely the technologically aided translation projects, the two participants who suggested excluding the part of *evaluation of the adopted CAT tools* from the projects attracted the researcher's attention:

[Quote 34]
"自己现在了解还不是很深入，就算做出评估，结果也不是很客观，因为本身就一知半解的。"
["*I'm afraid that without sound mastery of the tool, the evaluation I made may not be objective enough and won't do justice to the tool.*"]

[Quote 35]
"觉得这个评估意义不大，CAT 各自都有自己的特点，译员用过以后就会知道哪个最适合自己。评估更适合给软件开发人员进行培训。"
["*I don't think the evaluation part meaningful. The CAT tools are all different from each other. The users will naturally know which one fits their needs best after using them. Evaluation results should be used for training developers of the tools.*"]

From the above quotes, we can see that they either did not understand the purpose of the evaluation task or did not feel confident in making evaluation on their own. Again, regardless of the small number of the suggestions, the author found the message important and decided to place greater stress on the explanation about the function and importance of evaluation making in learning with the PBL approach.

6. 2. 4 A summary

Luckily enough, all the participants of this course happened to be well-motivated, possibly because the project was set in the small term when they had relatively more sufficient time at their disposal. Nevertheless, all the concerns and suggestions above point to the directions future research should address before the PBL approach can turn out desirable outcomes when implemented under different conditions.

6. 3   Findings of the second level: learning

This section will summarise the findings at the level of *learning*, namely the knowledge, skills and attitudes the participants have acquired in this course. Data analysed for this purpose include both the self-reported part, namely the responses to the second and fifth questions in the survey and information in the reflective reports concerning what has been learnt, and the factual data source of the *summative assessment documents*. The findings from the former were based on the content coded with *knowledge about CAT, MT and translation project, knowledge of CAT technologies operation* and *attitude towards CAT and CAT learning* as defined in Section 3. 3. 3. 3. 5. The findings from the latter were results of analysis in reference with the objects of assessment in terms of the ILOs.

6. 3. 1   Findings from the self-reported data

Generally speaking, in their reflections, the participants reported a great increase in their knowledge of CAT with only one exception who did not find any noticeable new learning. However, all admitted that after taking the course, they had been convinced of the importance of TT for professional translators, with 20 finding it indispensably important and 4 very important. In terms of attitude, 12 of the participants expressed greatly enhanced motivation to further study it and 9 moderately enhanced. 2 participants did not mention any change in their attitude and 1 reported, in sharp contrast, much weakened motivation to learn about CAT in the future. Quotes will be displayed and analysed in more detail below.

Firstly, about the exact acquisition from the course, the data covered almost all the ILOs anticipated by the tutor. The following quote is the most comprehensive comment:

［Quote 36］
"我在这门课里学到了:

(1) CAT 工具的应用

首先我们学习了 Google、BYU、语帆术语宝、diffen. org 等广义 CAT 软件的运用，然后学习了 Trados 的使用。无论是前者或后者，都是通过"自己寻找教学资料 — 在操作中学习 — 小组讨论交流，相互学习 — 班级内的交流，相互借鉴学习 — 总结"这个学习过程进行的。过程中遇上难题可以咨询老师和专业人士。

(2) independent-learning 的态度

从公选课的热门程度看来，授课老师受欢迎的原因除了给分高外，就是他们的教授称号。翻院教授级别的人不多（特别是在大一、大二），因此大家都希望在公选课能遇上'有水平'的教授们，期盼这些大师级人物能教会更多的知识。但是经常的情况是，即使上课的时候，教授们给学生传授了一些让学生耳目一新、甚至醍醐灌顶的信息，学生转头就忘了。这是因为学生并没有通过亲身试验，得到的并非是 first-hand 的信息，所以几乎所有的信息都没有内化储存。而课程的 independent-learning 则是 problem-driven 的学习模式，目的性强、操作性强，因此所获得的信息更容易转化为个人的知识。

(3) 小组合作

虽说翻院经常有 Presentation 需要小组的合作，但是那只是表面的小组合作。这种合作只是事前大致讨论一下主题，然后每个人选定子主题，各自做 PPT 或者准备展示，最后一天把 PPT 或展示合并起来，就成了一份粗糙的'小组合作成果'。而这一个课程的小组合作让我感受到了团队合作的精华。首先我们有明确的共同目标；再者，我们有一个做事有条不紊的组长；另外，团队里的每一位组员都对项目十分的负责、有热情、准备充分、及时地完成任务、对项目都有自己的贡献，从来没有出现一边倒的现象；最后，小组内的交流非常地积极、顺畅。因此我们一直以来的团队合作都十分的高效、顺畅、平衡、稳定。

本次课程的收获当然远不止这些，还包括翻译的前景等等。"

[" *I have learnt from this course:*
(1) CAT tools and their application
First we have learnt such broad-sense CAT tools as Google,
BYU, Yufan Termbase manager, diffen.org and their
application in practice. Then Trados, its operation.
Whichever tool we learnt, we learnt it via the same process-
'identifying on our own the learning resources — learning
by doing — learning from each other in groups through
discussion and communication — sharing learning
outcomes in class — summarization and reflection'. During
the learning process, the tutor and the invited experts were
always available for consultation.
(2) Independent-learning attitude
Judging by the popularity of those selective courses, most
teachers are popular with students and win high marks in
evaluation either because they usually grant high grades to
students or because they shine with the title of professor.
We don't have so many professors teaching us at SIS
(especially in the first and second years) and therefore all
look forward to selecting courses given by those 'erudite'
professors, expecting to learn more from them. However,
what normally turns out is the fact that however refreshing
and inspiring the information the professors convey in the
class, students just forget it instantly after they step out of
the classroom. That is because the students could not
assimilate the information without first-hand experience
and practice with it. Instead, with the PBL approach, we
had to learn independently, set very clear objectives, and
engage in ample first-hand practice. Knowledge acquired
this way could be consequently incorporated easily into our
existing knowledge system.
(3) Group work skills
In SIS we did a lot of presentations in groups, but the
cooperation in that group work was often quite superficial.

The so-called cooperation was no more than a brief discussion about the selection of a topic. Then when the topic was set and sub-topics divided, we just went back preparing our PPTs and presentations separately which would be pieced together on the last moment to form barely a work of 'group work cooperation'. Differently, in group study of this course I have witnessed the very essence of real cooperation: first we had a very clear common goal to work toward; then we had a leader who was very well-organized; what's more every member in the group were highly responsible and passionate, all fully prepared for their jobs and completed their share of work timely. Everybody made his contribution and there were no 'free riders' at all. Lastly, we communicated actively and smoothly within groups. Thanks to all this, our group had worked together efficiently and steadily.

Of course I have learnt much more. There is also the prospects of the translation industry, and so on and so forth."]

From this quote, we can see that the learner was not only aware of what she had learnt but also how she had learnt it. According to her report, more than successfully acquiring the required knowledge, she had clearly realized how the PBL approach had enabled her to learn as desired via a cyclic learning process featuring independent learning and study in small groups. Her case represents the ideal teaching outcome the author could expect.

What impressed the author most when reading the reports is that in summarizing what they had learnt most of the participants did not merely make a list of the tools they had mastered; rather some of them described in detail the learning process as shown in the above quote, some sorted out the major thread of the whole course, and some others reflected how the way the course was arranged had helped them learn beyond expectation. Another two quotes were copied here as examples with the reflective parts highlighted in boldness.

［Quote 37］
"课程是循着了解翻译的历史 —— MT 及搜索引擎 —— CAT 实战展开的，采用 Problem-based 的方法，其中穿插着密集的反思总结，这样的课程设置是很合理的，循序渐进且让同学去自主学习，确实能让学生学到很多。"
［"*The course follows the thread from history of translation to MT and search engines to first-hand experience with CAT. It adopted the PBL approach and weaved into it frequent exercises of reflections and summarisation. Very reasonably designed, the course enabled us to progress step by step and learn independently. As a result, the learners did learn a lot.*"］

［Quote 38］
"在接触这门课程之前，听过 Trados，一直觉得它很神秘。我一直对它充满了好奇。在这堂课上，没有接受更多的培训，在看完 11 个教学视频后，就开始用 trados。一边做项目，一边学习。在这个过程中，遇到了问题，就上网搜索或者咨询专家，慢慢地也掌握了 trados 的基本使用方法。现在回想起来，我觉得其实如果先从整体上能理解 Trados 的运行模式，记忆库、术语库、项目文件等概念，在实践中熟悉下各个按钮的功能，操作 Trados 并没有想象中那么难。"
［"*Before taking this course, I had heard about Trados and found it very mysterious. I had been quite curious about it. In this course, we did not receive much training (from the tutor); instead after watching 11 instruction videos we started to use Trados right away (in a project). We were learning to operate it while doing the project. Whenever confronted with problems, we searched online for solutions or went and consulted the experts. Gradually I came to grasp its basic functions. Now in retrospect, I realize that it is not so difficult as imagined to learn to operate Trados. What it takes is just to understand the working mechanism of Trados first, getting to the concepts of Translation Memory, Term Base and project file and then get familiar*

with the function of each button in practice."]

Though the two students worded differently, they both acknowledged the power of the reflective, independent learning characteristic of the PBL approach in getting them the expected learning outcomes, even something beyond their imagination.

Moreover, most participants noted down more specifically what they had learnt as the following examples show.

[Quote 39]
"在这门课程中，我接触和使用到了更多的翻译工具，包括：语料库，搜索引擎的高级搜索，不同功能偏向的字典，百科网站和其他的翻译资源。之后还学习了专业的 Trados 计算机辅助翻译工具，并实际地利用这个软件去完成一个项目。所有这些工具的习得都革新了我对翻译的理解。"
["*In this course, I had come to know and apply more translation tools, including corpora, search engines, e-dictionaries of different functionalities, encyclopedia websites and so on. Later I also learnt the specialized translation tool Trados and put it into real use to accomplish a translation project. Acquisition of these tools has renovated my conception of translation."*]

[Quote 40]
"翻译工具的使用，小组的合作，学习能力的提升，对于翻译产业的认识提升。"
["*(I have got) translation tools and their application, cooperation skills in group work, enhanced learning ability, improved understanding of the translation industry."*]

[Quote 41]
"除了 CAT 工具的使用外，还学到了下面这些东西：
1. 不能忽略一些看似简单的概念，比如说翻译文件的准备，深入去看也是一门大学问。
2. 团队成员紧密合作的必要条件是大家都认同团队的目标。

3. 早期的规划可以使我们更加高效地学习，比如刚开始时 Learning issues 的建立让我们的目标更加清楚。
4. 一些翻译的技巧，像是对英文长句的处理等等。
5. 小组成员都有自己出众的地方，从她们身上也学到了不少东西。"

["*Besides CAT tools, I have also learnt the following stuff:*

1. We should not neglect some seemingly simple concepts, such as preparation of translation documents. Put in perspective, it is an important skill.

2. A prerequisite to close cooperation in group work is a common goal for the group that is identified by each and every group member.

3. Early planning can enhance our learning efficiency. For example, the clarification of the learning issues make us clear about where we are going.

4. Some translation skills, like how to deal with long English sentences.

5. Distinctive qualities of other group members."]

More specifically, the tools the participants claimed they have learnt are listed below ranked with decreasing popularity measured by the number of the learners who mentioned them in their report. They are: SDL Trados 2011 (22), search engines (21), corpora (21), online encyclopedias (13), online dictionaries (12), online expert resources (11), Oriental Yaxin (10), E-libraries (6), and online MT (3).

Secondly, as to the change in attitude toward CAT and CAT learning, 21 participants found themselves more motivated to further study CAT and apply it in their practice. Some typical reasons are listed below:

Some participants became more interested to know more after overcoming art students' stereotyped fear of technology learning.

[Quote 42]
"通过运用觉得 CAT 也不是那么遥不可及的工具。"
["*Having applied it to real use, now I find CAT is not that inaccessible to me.*"]

［Quote 43］
"发现了 CAT 的好处，也发现了其他有用有趣的网站，希望能积累更多。"
［*"Now that I have found the advantages of CAT and many other interesting websites, I want to know more."*］

Some still had some reservations about CAT, but sensed the trend of computerization of the translation industry and therefore showed a moderate desire to master basic skills in this regard.

［Quote 44］
"以前对于 CAT 是蛮好奇的，接触以后感觉技术层面的东西很多，核心的东西还是人的语言能力，以后还是很有必要学习翻译工具，但是不能全依赖工具，本身的语言能力还是很关键的。会继续了解学习 TRADOS，但也许不会主动去多学其他 CAT 软件……"
［*"I had been curious about CAT. After learning it I found that the tools in themselves are much too technical. The very key to successful translation is still the translator's language competence. So it is necessary to learn about translation tools, but we cannot count on it entirely. Development of our own language competence is still very critical. I will go on learning Trados, but probably no others."*］

Many more became motivated upon realizing the power of CAT in promoting their proficiency and quality of work. A few were even fascinated by the potential advantages CAT may bring them in their career development and thus showed strong eagerness to be specialized in this field. Following are some examples.

［Quote 45］
"本身喜欢翻译，但是之前做翻译都是传统的方法，效率低。这门课让我发现了世外桃源，发现了 CAT 这个便捷翻译的途径。"
［*"I am fond of doing translation but have done it in the traditional way which is very inefficient. This course opens to me a whole new world, bringing CAT to me, this highly*

197

efficient approach."]

[Quote 46]
"因为掌握更多的技能，翻译的质量与效率就会得到很大的提高，对以后可能的翻译 career 也提供更多的机会。"
["*With this additional skill (i.e. mastery of CAT), the quality and efficiency of my translation will be greatly enhanced. Then I can seize more opportunities in future career."*]

[Quote 47]
"翻译工具使得翻译效率提高，对于未来的职业发展会有很大的帮助。"
["*Mastery of translation tools can promote work efficiency, which provides us an edge in our career development."*]

[Quote 48]
"有了明确的概念，甚至有了要做一名技术宅的想法，这技术化应该也是一个趋势吧。"
["*Having understood what CAT is all about, I realized CAT is an irreversible trend and even dreaming of becoming a CAT aficionado."*]

Only one participant felt less motivated than before the course, saying that:

[Quote 49]
"我学习 CAT 的动力没有以前强了。选择这门课的时候，以为学习到了翻译工具能够减轻人力负担，能够在短时间内学习很多资源和知识，课程结束，发现工具只是工具，代替不了人工，只有不懈怠地不断提升综合素质，才能跟得上行业发展。"
["*I am now less motivated to learn CAT than before. I wanted to learn CAT tools from this course hoping that I could obtain a lot more resources and knowledge which*

> *would reduce my workload. Now the course is over. I find*
> *tools are tools. They can never replace human translators.*
> *What I should do is still to upgrade my overall competence*
> *in order to keep up with the development of the*
> *industry."*]

A close reading of this comment, however, shows us lack of logic between her statement of lessened motivation and the reasons she gave for it. The frustration might not result from the fact that CAT failed to meet her needs; rather it might be due to her misunderstanding the role of CAT tools in translation process which laid her hope on the wrong place.

All in all, to extract the knowledge reported to have been acquired at the level of learning, we can say that this course has produced highly satisfactory ILOs at this level. As evidence shows, when rating their own ability to apply the learnt tools, all the participants believed they had met the minimal requirements of the course, with 11 of them thinking that they had known how to apply them but not skillfully enough and 12 of them feeling quite confident in applying a variety of tools to different needs in practice.

6. 3. 2   Findings from the factual data

Although quite affirmative, the self-reported learning outcomes are not as reliable as those shown in their actual performance on the assessments. The following part will then summarise the observable learning outcomes of the participants gathered in the summative assessments and the individual journals as well as the group study records. As shown in Table 3. 4 in Section 3. 3. 3. 3. 5, aligned with the level of learning are ILOs 1, 2, 3, 4, 6 categorized by theme into Codes 2, 3 and 4. Therefore, all the assessment tasks involving these ILOs (see Table 3. 3 in Section 3. 3. 3. 3. 3 and Table 5. 12 in Section 5. 2. 5. 2 for details) except the individual reflective reports were examined and coded in search of evidence for the leaners' changes at this level.

6. 3. 2. 1 Findings from the summative assessment documents

This part of data shows us that all the groups had successfully accomplished the tasks using both broad-sense and narrow-sense CAT tools and newly acquired knowledge about translation project management.

In correspondence with Code 2, firstly 13 essay questions (Appendix 16) were designed to test students' knowledge about the basic yet critical concepts concerning CAT technology and its application with 3 questions compulsory and the rest optional. In additional to the 3 compulsory questions, the learners were required to choose 2 more questions out of the remaining 11 ones to answer without referring to any external help. That is, every learner was supposed to answer 5 questions in all with each scored 20 points. The final score of this task is supposed to provide evidence for the learners' knowledge about CAT, MT and translation project (Code 2). Then the final mean score is 92 which indicates to us a quite satisfying result of the learners' acquisition of the knowledge part of the course.

Moreover, special attention was paid to how CAT tools were adopted in the work flow design for the translation project based on the assumption that reasonable adoption of CAT tools evidenced familiarity of the users with their functions. It was found that each group was highly aware of the basic functions of the adopted narrow-sense CAT system, in this case Oriental Yaxin and SDL Trados 2011, such as how to make TM on WinAlign, how to extract terms on Multiterm or Yaxin, and how to translate and revise using the workbench. All the groups also included in their plan recommendable broad-sense CAT resources for reference. For example, group one listed explicitly the CAT tools and the purposes they were adopted for as in Table 6. 1. In addition, all the groups had gone successfully through the whole translation project process submitting all the required documents (See Table 3. 3). Judging by the completeness and quality of the documents, we could conclude that all the participants were aware of the basic procedures of a translation project and capable of managing it in a cooperative way with technological aids. Yet as reported in the last reflective reports, without too much experience with such a technologically enabled large scale project, the learners found translation quality assurance, faculty evaluation, and risk control relatively more difficult to handle and therefore in these aspects followed the suits from the sample cases from the tutor rather than come up with original ideas.

Table 6.1   CAT tools adopted by Group 1

Tools adopted	Purposes
Trados 2011 Studio	生成记忆库并翻译、质检校对功能、生成项目并传递以便提升翻译速度和翻译质量
Trados 2011 Multiterm	生成术语库
Microsoft Word 2011	编辑排版目标翻译文件
Solid Convert PDF Pro	PDF格式文件转为Word文件
Google	查找专业背景知识和专业表达
灵格斯词典 (Lingoes)	查找单词，便于验证术语库及翻译

In addition, in correspondence with Code 3, it was found that when describing the technological aids during the work flow design, no group delved into details of operation regardless of the tutor's requirement for it, making it unclear to the tutor how exactly the tools and resources were used. This suggests to the author that the leaners might either lack enough operational knowledge at the time of planning about the tools or feel not confident in their plan in fear of possible errors that would lead to some deduction in their grade. The guess turned out true when it was found later when asked to revise their project plan based on the lessons they had got during its implementation, some groups were able to add desired details of CAT adoption into the workflow, as shown below in the example from group one (see Figure 6.2).

Besides, among the available functions of Trados, the QA functionality was found to be the least mentioned and applied one. Some groups admitted in their final report that unfamiliar with the self-defined QA module, they relied more on themselves to proofread the final translation, missing even the step of automatic checkup altogether. Questions from the learners on QQ rose abruptly during

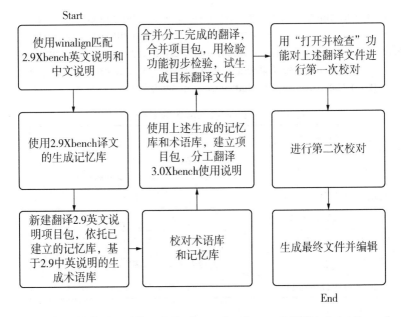

Figure 6.2   Revised workflow design integrating the use of CAT tools by Group 1

the last a few days when it came to the last stage of the project, namely the review and editing of the final product. The advanced functions of Trados, including reviewing, QA, and text output, seem to be very difficult to master, especially when the format of the original text is complicated. Therefore, in terms of Code 3, we have observed less satisfying results than Code 2.

Lastly, the participants were at the end of the course required to hand in their portfolios which should contain all their self-initiated efforts during the course. Now these portfolios are read closely as part of evidence for the participants' attitude toward their learning. It is found that although the majority of the portfolios are mere copies of the learning resources and extra readings the tutor provided during the course to facilitate their learning, every participant included in their portfolio something new, if not as substantial as expected. The more or less initiated efforts to learn more than the tutor had taught, especially on such a tight schedule, is a good indicator of the participants' positive attitude toward learning of

CAT. To make their efforts rewarding, upon finishing reading them, the author picked out the new additions, put them into categories and shared them via the Internet with the whole class (see Figure 6. 3).

词汇术语	2013/10/13 11:59
翻译技能	2013/10/13 12:01
翻译技术理论	2013/10/13 11:58
广义CAT	2013/10/16 22:03
行业信息	2013/10/16 22:03
其它	2013/10/16 22:04
双语 记忆库资源	2013/10/13 11:57
狭义CAT	2013/10/16 21:57
写作	2013/10/16 21:58

Figure 6. 3   Screenshot of the new additions to the learning resources by the participants

In sum, as this part of factual data shows, sufficient evidence has been found that the learners had all mastered the basic functions of both board-sense and narrow-sense CAT tools, although some clues presented themselves for the learners' less satisfying command of such advanced functions of Trados as reviewing and QA. Moreover some steps in the translation project were dealt with less desirably, such as translation quality assurance, faculty evaluation and risk control. Nevertheless, the successful completion of the project in a nearly professional way by each of the group convinces us of the fact that they have achieved, to a great degree, the ILOs at the level of learning.

6. 3. 2. 2   Findings from the journals and records

Comparatively speaking, the individual journals and the group study records provide more details of what and how the learners had learnt.

A scan throughout all the journals and records reassured us that each and every of the participants followed strictly through the learning schedule. The limited space here does not allow the author to display all these learning experiences, so she picked out the participant with the smallest number of journals (i.e. 洪奭 Hong Shi) and tried to see how the learner had learnt. The result amazed the author in that the learner actually showed a great amount of self-directed learning and independent thinking following yet not confined to the broad schedule of the course. She did not note down her learning experience every day as required yet every time she kept the journal, she did it very seriously. This is reflected not only in the length of her journals, which are normally longer than 2 pages, but also in the content. The most surprising finding is that inspired by the reading materials offered to them in the first class, she expressed a strong interest to know more about CAT in the first journal, and in later journals, many instances were noted down of how she thought actively (or sometimes could not help thinking) about the functions and the prospect of CAT technology in everyday life whenever coming across incidents related to translation (see an example in her journal on August 22 attached in Appendix 17).

Both the way the journals were kept and the time amount reported by the learners in the journals were at first taken as indicators for the attitude of the learners towards their study in the course. The underlying assumption is that the more detailed the journals are and the longer time the learners spent on this course, the more positive their attitude. A general view of the content of the journals however seemed to be a bit discouraging to the researcher as most of the participants seemed to have just scribbled down their learning experience without enough desirable details. The author therefore selected a few learners who she found to be very active and conscientious during the course and went to read their journals in depth, trying to see whether the way of journal keeping has anything to do with the learner's attitude. The result was no. It seemed that the way of journal keeping is more a result of a personal style. No regular patterning was found to bear any correlative relationship with the attitude of the learners. Then this standard was dropped. A closer look at the statistics was then carried out. The result shows us that the requirement for keeping journals on daily basis proves to be highly demanding as only 5 out of the 24 participants stuck it out from August 14 to September 5 without missing a single day. The minimum is 14 journals only. But regardless of the missing journals, the overall amount of time each of the participants spent on the course was surprisingly huge, with an average of 188. 22 minutes per day and the most devoted one 315. 68 minutes per day. Such a figure manifests no doubt the learners' great devotion to the course, yet also concerned the author with the idea that such heavy workload will make it impossible for "normal" terms.

For now, it should be safe to say that consistent with the self-reported enhanced motivation to learn CAT, the learners did devote greatly their time and energy to learning in the course.

6. 4   Findings of the third level: behaviour

Clues about the changes the course has caused in the learners' subsequent behaviour were also found in this part of data of both self-reported and factual kinds.

First the reflections on changes in one's behavior were addressed by Questions 5 and 9 of the questionnaire survey. Among the findings, responses to Question 5 centre mostly on their changed attitude toward study and the different learning habits as the result of this course. Responses to Question 9, however, reveal to us more of their changes in using the CAT tools in their later learning and practice. Then again seeing the poorer reliability of the self-reported results, more evidence was searched for in the factual data. The factual data included in this part are part of the summative assessment documents related with ILOs categorized into the third level of behaviour (see Table 3. 3 in Section 3. 3. 3. 3. 3 and Table 3. 4 in Section 3. 3. 3. 3. 5), the observational sheet and the tutor journals. They were coded in search of the themes 5 to 9 as specified in Table 3. 4 in Section 3. 3. 3. 3. 5 and reported below in detail.

6. 4. 1   *Findings from the self-reported data*

6. 4. 1. 1   Findings from Question 5

Among the responses to Question 5, we found that almost all

the participants mentioned that their ability to learn independently had been greatly improved. More details of how can be seen in the following quotes.

［Quote 50］
"通过帮助别人解答问题发现了自己知识结构上的缺陷。现在学会了自主学习，端正了学习态度，觉得这样的学习方法让自己收获很大。"
［"*I have detected weaknesses in my knowledge structure through solving problems for others. Now that I have learnt how to study independently and adjusted my attitude toward study, I have benefited a lot from this change.*"］

［Quote 51］
"有的。1. 对老师的依赖性降低了，摆正了自己作为学生的心态，我的学习和教育应该由我主导。2. 有效利用资源和工具，多查多问。"
［"*Yes. 1. I am not so reliant on teachers as before. I have now changed my attitude toward study, realizing that learning and education are better self-directed. 2. Now I can make good use of different resources and tools, and are willing to consult others, in search of possible solutions.*"］

［Quote 52］
"有。不再依赖于课堂，而是将学习任务分配到每一天。"
［"*Yes. I am no more dependent on the learning in classrooms. And I can make more reasonable plan for my study every day.*"］

［Quote 53］
"有。培养了自己独立思考的能力，现在遇到困难和问题时，首先思考自己可以如何独立解决而不是一问地询问他人。"
［"*Yes. My dependent thinking ability has been greatly improved. Now when I encounter difficulties and problems, I would start with thinking how I can solve them in my*

own way instead of asking for others' opinions."]

[Quote 54]
"我会主动在小组内提出问题，然后和组员们进行讨论，自己去摸索答案，而不是一味地遇到问题就没头没脑地冲进去回答。"
[*" Now I would like, more willingly, to ask questions and discuss problems with other members, searching for answers ourselves, instead of rushing for an answer without too much thinking."*]

[Quote 55]
"这个项目完成下来，自己的耐心也变好了很多，更乐意去接受别人的意见。"
[*" Upon the completion of this project, I found myself much more patient than before and more willing to accept others' opinions."*]

[Quote 56]
"在之后的课程中，我感觉我更加会自己发现问题解决问题了，思维发生了改变。"
[*" As the course progresses, I found myself more capable of identifying and solving problems. The way of thinking has changed."*]

[Quote 57]
"研究表明，当一件事情每天都做，21 天后，就会形成习惯。所以，上完这一个月的课程后，得益于每天的 study journal，我发现自己已经养成了每天 reflect 的习惯。"
[*" Research shows that it will grow into a habit if something is done every day. That is why I found myself used to reflecting every day thanks to the assignment of keeping journals during that month in the course."*]

[Quote 58]
"以前只是做好老师说的，自己扎实点，现在感觉我会

用批判的思维去对待自己的功课，自己想是不是真的要按照老师说的做，有没有别的解决方法；遇到问题的时候也更加会想自己解决，不是一上来就问人；而且感觉解决问题的能力也增强了。"

["*I used to learn strictly as teachers required. Now I am more critical. When doing an assignment, I would not simply follow the teacher's instruction. I would ask myself if this is the only way to do it. When faced with a problem, I would not count on others for solutions; rather I would like more eagerly to solve it myself. What's more, my ability to solve problems has been greatly enhanced.*"]

[Quote 59]
"有变化。对于学习，以前比较缺少自己的思考，也不是经常提问。学习完课程后，我感觉，学习的重点不是看书，背书，更重要的是提问和反思。现在做其他的事情的时候，也会习惯性地先提问和思考，做完后，也会停下来反思一下。"

["*Yes. I had not very often thought independently nor asked my own questions when studying. After the course was over, I realized that it is of no use to learn by rote. It is more important to have one's own questions and reflect on what one has learnt. Now even when I am doing other jobs, I would habitually ask questions and think about them myself. After finishing them I would stop to look back at them for a while.*"]

[Quote 60]
"有变化：学会分享成果与问题。"
["*Yes: I am now willing to share my findings and problems with others.*"]

To sum up, during or after the course, the learners have grown the habit of asking their own questions when met with new problems, become more courageous and skillful to pursue their own solutions, and shown more willingness to cooperate and share fruits with other, and high awareness of reflecting on their performance. All this is a good sign of the effectiveness of the PBL approach in promoting students' independent learning ability.

6. 4. 1. 2   Findings from Question 9

Comparatively, responses to Question 9 were expected to be more reliable since the question was delivered to the participants via email one year after the course was over and with such a long interval in between, there would presumably be more opportunities for the changes to emerge. Analysis of data from this part proved the expectation to be true. The responses collected do show us more evidence for changes in both the way they deal with problems in and their actual application of CAT tools (in broad or narrow sense) to their later learning and practice.

Before moving on to the exact changes, what is worth mentioning here is that, after one whole year, in responding to Question 9, 16 participants expressed openly their gratefulness for having taken this course and stated that they could still remember clearly what they had learnt from this course. A good example is as follows:

[Quote 61]
"去年的计算机辅助翻译课程，算得上是所有翻译课程中对我意义最大，也是我投入最大的一门。在此我必须感谢骆老师，您为我打开了翻译的新世界，也通过您精心设计的高质量的 PBL 模式让我发自内心地爱上了 CAT。关于具体的课程内容，我想我还是能说出一二的。首先我们以祖孙三代的翻译故事为导入，探讨了翻译从旧时代的一笔一词典到如今专业化、产业化、项目化、信息化。（这也是几个 activities 里记忆比较深的一个）然后我们一方面学习了各种信息检索技巧，另一方面从对齐入手学习雅信和 trados 的使用。课程后期我们进行了 Xbench 使用手册 4.0 翻译项目。不足 2 周，4 个女生，实打实的挑战。但在这理论实践相结合的过程中，深刻体会了翻译项目管理和人机结合进行翻译的魅力。对于翻译行业未来趋势的展望也给了我的思考和信心。"
["*The CAT course of last year should be the most*

meaningful course of translation for me and the course I had been most devoted to. For this, I have to extend my gratitude to Ms. Luo for opening a whole new world of translation to me and getting me in love with CAT by the well-designed PBL approach to the course. I think I can still name the specific content of this course. First, we began with making a story of three generations of translators, by which to explore the development of translation from the traditional individually based activity with pen and dictionary to today's highly professionalized, industrialized, and computerized project-based activity (which is one of the activities I remembered best). After that we learnt, on the one hand, different skills of information searching, and, on the other, how to operate Oriental Yaxin and Trados, starting from the function of alignment. Later in the course we accomplished a project of translating Xbench User Guide 4.0 using the CAT tools. In less than 2 weeks, for us, the 4 girls, it was a real challenge. But it was during the process when the theories were put into real use that we were deeply impressed by the glamour of translation management and coordinated translation aided by machine. Lastly the introduction to the prospect of the translation industry provoked more of my thought and gave me more hope for the future."]

What impressed and touched the author most is that one participant could even recall in great detail how the PBL approach was introduced and implemented in the course:

[Quote 62]
"对于课程的内容，我大概还记得，因为是由我们自己决定的。一开始是介绍课程的内容：是由我们自己决定怎么学！老师只是担当引路人的作用来的。然后介绍课程所采用的教学法的根据，每节课后有相关的文献给我们自己阅读。后来是分小组确定我们的学习目标和学习方法～后续的课程有很多小组的工作，一步步引导到我们的项目：自学 CAT 工具，并利用其来翻译说明书。我

们每天要写 study journal 来记录自己的学习历程，组长
最后把小组的学习成果都汇总出来上交。"

[*"As to the content of the course, I could still remember
because it was decided by ourselves. At the very beginning
the tutor introduced the course to us, telling us that we
were gonna decide how to learn ourselves! The tutor would
only be a guide. Later she introduced to us the underlying
philosophies of the approach. We would have some
literature to read on our own after each class. Then we were
divided into groups and discussed to determine our learning
issues and how we were going to learn them. Step by step
we were guided to learn and lastly to the project in which
we had to learn to use CAT tools and apply them to
translating a user guide. Besides, we had to keep study
journals every day recording our learning experience. At
the end of the course, the group leader had to collect all the
learning documents and handed them to the tutor."*]

Then as to the changes, first, among the 21 responses, 8 showed
evidence for changes in the way of thinking in their later learning and
even life experience other than translation. For example, some
participants found themselves more active to use a variety of tools in
search of solutions to problems:

[Quote 63]
"这个课程一个月的学习十分充实，一次性接受了很多
新事物。印象最深的应该是有了使用工具的概念和意
识，在遇到问题时能主动查找利用资源去解决，思维更
加严谨。"

[*" That was a very busy month. I had learnt a lot new stuff
in this single course. What impressed me most should be
the consciousness of the value and the necessity of
instruments. Now met with problems, I can go and find
resources actively and use them to figure out solutions.
And I can now think more logically."*]

Besides, many more mentioned they can think more logically (as
in the above quote) and react more calmly when in difficulty. For

example,

[Quote 64]
"小组合作的经历是我第一次在翻译、笔译项目中的小组合作。这个小组合作很彻底，让我之后笔译学习中进行小组合作时感觉思路很清晰，比其他同学更能 hold住。"
["*The group work in that course was the first time I translated in a group-based project. We worked as a real group. (Thanks to the experience) later when I learnt to translate in groups I could always keep myself cool-minded and hold pressure better than my group members.*"]

[Quote 65]
"一个月来，作为最艰难的小组的成员，本来技术不过关，还兼顾着学生组织的工作，任务量觉得很大，所以后来学期的课程，任务再多，其他同学抱怨再大，我都会觉得，跟小学期那个月比起来，真的没什么，就特别自我满足，所以后面的学期就自我感觉很轻松！"
["*During that month, as a member of the group in greatest difficulty, I felt enormous pressure due to, on the one hand, my poor mastery of the tools and, on the other, the extra work I undertook in other student organizations. With this experience, later in other courses, I would feel nothing however heavy the workload was. They were no big deal at all compared with what we had experienced that month. Therefore, hearing my classmates complaining, I felt a great sense of achievement. Learning was quite easy for me in the following term!*"]

In addition to the resulting stronger willpower to meet challenges, better cooperation skills, greater independent learning ability and transferability of the acquired problem solving skills were also mentioned widely in the responses. Some are quoted below.

[Quote 66]
"在大三的笔译项目中（两次），和多次合译中，注意考虑到之前犯错的地方，例如，在多项选择中更果断，犯

错不要紧，沟通最重要。其实这算是思维的一种转变
吧。"
[*" In the two translation projects in the junior year and
many more cooperative translation tasks, I benefited from
the lessons I had learnt from the previous experience (in the
course), such as 'be more decisive when making choices',
'fear not making mistakes', and 'communication matters'.
This is owing to the change in the way of thinking, isn't
it? "*]

[Quote 67]
"我从课程中学习的技能和知识帮助到我的学习和实践。
我觉得我学习到的更多是学习的方式，另一种接受知识
的方式，自己发现问题解决问题的思维，以及自信心。
虽然具体怎么用 Trados、雅信等工具，可能后来还未用
得上。但是因为在课程中我们是小组的形式自学并完成
了任务，这使我觉得，有一天我要用的时候，我会记
得；即使我不记得，我也会有能力重新找到我自己需要
的知识并解决问题，因为我曾经这么做过。后来在学习
字幕制作、嵌入字幕等的过程中，我觉得自己自信心更
强，不会有畏惧心理而且上手也比较快。"
[*" The skills and knowledge I had acquired from the course
have helped me in my later learning and practice. I think
what I have got was more about the learning approach,
another way of absorbing knowledge, a way of finding and
solving problems, and more confidence. Although I have
not got chances to apply Trados and Oriental Yaxin in
practice, the fact that we had learnt how to use them and
had accomplished a task with them in groups on our own
made me believe that I would be able to pick them up
whenever I need to use them and that even if I could not
remember exactly how to operate them, I could recollect
the knowledge and solve the problem. It is all because I
have done it before. It actually did happen when I learnt
subtitling later. During the process of learning, I found
myself more confident, without much fear, and could learn*]

more quickly."]

［Quote 68］
"甚至推而广之，我在其他领域的学习，和日常生活中，也受到了这次课程的影响。我认为课程所运用的学习方法培养了我的一种新思维：我在口译学习，甚至是生活中如烹饪学习、修理电脑等的时候，都可以运用这钟思维：自己发掘问题，运用工具收集资料，解决问题。"
［"*In a broader context, I could even feel the benefits of the course extend to fields other than learning, to my everyday life. I think the learning approach adopted by the course got me to think differently, which applied to my study of interpreting and in life like in learning cooking, learning to fix a computer etc. That is, I would find my problem, collect information using necessary tools for solutions to it.*"]

However, as to the actual application of the acquired CAT tools, the findings are not as promising. Comparatively speaking, while the broad-sense CAT tools and resources have been more or less used by all the 21 respondents of the question, only 10 of them have used the narrow-sense CAT tools afterwards occasionally. Most of the respondents admitted that these narrow-sense CAT tools are too powerful to be needed by learners at their level. See examples below:

［Quote 69］
"感觉在平时做翻译的时候，还是比较少需要用到翻译软件工具的，反倒是一些搜索信息的方法更加实用一些。"
［"*I found that we didn't really need to uses the specialized CAT systems in our everyday translation practice; rather some information search methods are more helpful.*"]

［Quote 70］
"然而，很遗憾，由于所接触材料篇幅和题材原因，自课程结束至今，并未使用过 Trados 及该系列软件进行翻译。"

[*"It's a pity, however, that since the course was over, I have never used Trados or similar CAT systems to translate, due to the genre and the volume of the translated texts."*]

[Quote 71]
"专门的 CAT 工具在我这一年的学习中基本没有用到，为数不多的接触是在学习法律翻译时，那里面有很多重复，所以当时找了一些平行文本，使用 Trados 完成了笔译老师布置的作业。（当时是翻保密协议，网上有很多合同的平行文本，有大量的重复，心中一喜。但是毕竟平行文本和老师布置的作业不是完全贴合的，因为我自己懒得去一行行对，所以就用了 Trados，用 Trados 来检查不同点会省事不少。）"
[*"I have almost never used the specialized CAT tools in my study of the past year. The only exception was the time when we learnt legal translation. I found there was a lot of repetition in the legal text the teacher assigned us, so I searched for some parallel texts, fed them into Trados and did the translation on it. (We were then asked to translate a confidentiality agreement. I gleefully found a large amount of parallel texts with a lot of repetition with the task. Of course these parallel texts could not be exactly the same with the one to be translated and I was too lazy to align them with my naked eyes. Here was where Trados came in! It saved me a lot of time checking the inconsistence between the texts."*]

[Quote 72]
"Trados 用得不多，但对于熟悉其他 CAT 工具挺有帮助。之后有接触到其他的 CAT 工具（主要是翻译北大那个 MOOC 课程时接触到的在线 CAT 工具），慢慢发现 CAT 工具其实都差不多，因为比较详细学过 Trados，能较快了解每种 CAT 工具的基本功能。"
[*"I did not use Trados that much later but my knowledge about it made it easier for me to learn about other CAT*

tools. Later I have come across some other CAT tools (mainly when I was translating the MOOC course for Beijing University) and came to realize that there is a great degree of similarity between different systems. As I have learnt Trados in depth, it was much easier for me to understand the basic functions of those CAT tools."]

［Quote 73］
"由于加入了 CATwork Studio，使用翻译软件次数还是挺多的，比如说 MOOC 字幕翻译中用到了 google translation toolkit，平时也有继续学习 Trados 等软件，只是平时课堂学习和课后个人翻译实践中用到的情况就比较少了。据很多同学反映，他们在上完 CAT 课后都卸装了翻译软件。"
[" As I was recruited into the CAT work Studio, I've got many more chances to use specialized CAT tools. For example, I used Google Translation Toolkit when translating the MOOC course and kept learning Trados. Yet in everyday exercises and personal translation practice there wasn't much need to use them. As I was told, many of our classmates uninstalled the software right after the course was over."]

As shown in the above quotes, as the CAT tools are not applicable to the assignments from their translation courses which tended to be short literary texts, the participants did not have many chances to use them. Yet what is encouraging to the author is the finding that whenever the need arises, the learner knew how to apply the tools, though for limited purposes. Besides, the fact that some learners found their knowledge about Trados could be transferred when learning new tools shows the author the strong continuous learning ability which is also one of the major ILOs of the course.

To conclude, the reports from the participants showed us very strong positive evidence for their changed behaviour as the result of the course with the PBL approach. Nevertheless, this self-reported perception is not ideally reliable unless supported by factual data. More factual data will therefore be displayed and analysed in the next section.

6. 4. 2   Findings from the factual data

As stated earlier, factual data used to explore the learners' changes in behaviour are the related summative assessment documents, the tutor journals and the observational sheets. What is worth noting here is that as the observer unfortunately fell ill in the middle of the project and had to take a sick leave; only 7 meetings were observed, which could hardly tell the whole story of the classroom proceeding. Yet what was lucky is that the observations missed only the middle few days. Then with the beginning and the ending of the classroom tutorials recorded, the progress of the students can nevertheless be possibly noticed. Data of these three sources will be integrated below serving together to reveal the leaners' changes in their behaviour as a result of the course.

6. 4. 2. 1   Positive findings

First, the tutor journals were examined with special attention paid to the noted advancements in the learners' behaviour. Among the 10 journals altogether, it is quite clear that the tutor felt very touched by the learners' cooperation and devotion to the course especially in the later two weeks when the project was ongoing. The advancements often mentioned chronologically in the journals are more positive attitude toward CAT learning, greater willingness to study independently, more critical reaction to different opinions, more skills to cooperate with (and tolerate) group members, more resourceful broad-sense CAT application, richer knowledge of narrow-sense CAT application, ability to apply acquired knowledge in the translation projects, which covered themes 5, 8, 9 and 10.

Some of these changes were also noticed in the classroom observations, among which progress of the themes of 9 and 10 is the most obvious.

For example in the observational sheets of August 19 and September 2, the observer noted down the higher motivation to learn and exchange ideas:

［August 19］
"学生都能可以而且很主动的回答问题。"
［" *Students are able and motivated to answer questions.* "］

［September 2］
"学生比上学期传统教学法课上更加主动，参与性更强。"
［" *Students are more motivated and engaged than last term in the traditional class.* "］

Besides, the observations made on August 21, August 28, August 30 and September 2 tracked the progress in the learners' awareness and skills of problem solving.

［August 21］
"Group 2 那个组探讨得好深入！"
［" *Group 2 probed so deep into the problems!* "］

［August 28］
"有些同学在提问和解决问题时，只抓细节钻牛角尖了，没有意识到 big picture。"
［" *Some learners liked splitting hairs. They went too deep into details to see the big picture.* "］

［August 30］
"大家在摸索中能总结问题和应对方法……大家发现问题的敏锐度好高！"
［" *They were able to reflect upon the problem and find solutions through independent exploration… They were very sensitive to problems!* "］

［September 2］
"同学们已经有独立解决问题和讨论解决问题的意识和能力。"
［" *The learners have got not only the awareness of discussing problems on their own but the ability to solve them independently.* "］

As seen in the above comments, some of the learners showed potential for such explorative way of learning at the beginning of the course. As time went by, all the participants seemed to learn how to identify problems and explore solutions to them independently although on the way to the end some learners encountered some difficulty (as observed on August 28).

The other thing that impressed the observer was the warm and cooperative learning atmosphere.

[September 2]
"各组间会相互帮助，乐于解决他组的问题。"
[*"The groups all felt like helping each other with their problems."*]

[September 4]
"大多同学都乐于互相帮助。"
[*"Most of the learners are willing to help others."*]

Based on the reflections of the learners reported earlier in this chapter, we can see that the month of learning in groups makes them see the superior qualities of their classmates and realize the unparalleled power of the collective efforts. Willingness to help others would be a natural consequence from this awareness.

Besides, the process management and evaluation files the learners were required to include in the assessment documents provide some evidence for learners' accomplishment of ILO 11. It was found that with sufficient guidance from the tutor and sample cases to follow, the learners had all learnt to take effective measures for process control and staff evaluation. In the end, 4 out of the 5 groups accomplished the project task ahead of the schedule with Group 4 finishing all their work 3 days earlier than the deadline which shows at least partly that they have monitored their projects quite effectively. As to the results of the staff evaluation, they were kept open to every participant for the sake of fairness. Different than the previous years when the author would receive more or less complaints about the biased comments, she heard none this time. An example of how they evaluated their group members (see Table 6-2 below) might speak for itself as to the participants' ability to assess the performance of their own.

Table 6. 2　A sample staff evaluation sheet from Group 4

组员评分表				
评分内容				
项目前期	按时参加开工会，会上各种积极发言（10）			
	按时完成译前准备工作（10）			
	按时参加项目相关培训（10）			
项目期间	按时提交个人工作进度表（10）			
	严格根据进度完成任务（10）			
	认真负责，保持与各成员的沟通顺畅（10）			
	积极为遇到的问题提供解决意见（10）			
项目后期	完整保存、提交项目过程文件（10）			
	按时提交个人项目总结（10）			
	按时完成项目后期分配到的工作（10）			
总分（100）				

6. 4. 2. 2　Negative findings

Regardless the posifive findings mentioned above, there are also some behaviours of the learners that were found less satisfying. The most obvious ones in the three data sources will be presented below.

Firstly although in both the self-reported and the factual data, the learners' motivation for their study were shown to be greatly enhanced, their capability of independent learning was yet to be improved. Evidence for this conclusion is the observation noted in the tutor journal that the observed active questions were still undesirably few. The best example in this regard is the fact that the two experts the tutor invited to join the course were very seldom consulted, let alone challenged. While among the suggestions the learners made in their individual reflective reports many mentioned the hope of getting to know more real cases of translation project management, none actually asked the two experts from the "real world" about it at all.

Also related to the ability to learn independently, it was found, in the tutor journals, that the learners still had difficulty in identifying their own problems or needs. As part of the PBL approach, the learners were asked to identify learning issues at the beginning of the course under the tutor's guidance. Yet as noted in the journals, all the learners felt very much frustrated, knowing not at all what to do at the very beginning. Seeing this situation, the tutor slowed down and adjusted her plan, prolonging this 2-day task through a week and providing additional reading material and tutor guidance. But the result was still not satisfactory and 2 of the 5 groups kept complaining about their desperation about the task. Nevertheless finally the 5 groups all managed to decide their own learning issues, which were allowed to be modified later as their knowledge increased. The same thing occurred later when they began to learn the basic operation of Trados. To make the learning process more fun, the tutor designed an activity for them to challenge each other with questions about the operation skills, some real ones for seeking help and some faked for testing their opponents' learning effects. Yet again when asked for how they felt about this design most of the learners suggested deleting it altogether with the reason mentioned the most being they did not know what to ask or they did not think the questions they had asked meaningful at all. So these two incidents suggested to the author that the learners were still poor at identifying problems of their own, which is actually a critical starting point for any effective self-directed learning. The tutor's perception seems to contradict with the observer's comments in the above section on the learners' remarkable ability to find and solve problems themselves. Yet a second thought would make this understandable: The tutor and the observer, a student, may very reasonably have different criteria to judge the ability against. The different conclusions are therefore inevitable. Yet one thing is sure. That is, the fact that the learners did complete all the tasks successfully within such a tight schedule does show us the general improved ability in problem solving.

The second noticeable problem with the learners is that they were still less critical to different opinions than they themselves had thought. First of all, according to the tutor journal, very few students could actually discuss with the tutor at the equal footing. They might feel unable or unwilling to refute the tutor's views, turning discussion between the tutor and the group into a mini lecture (see [Quote 15] in Section 6.2.2).

As to the operation skills of the CAT tools, the observer noted on August 28 that no group had actually applied the CAT tools throughout the whole project as alleged. The author could not make any sensible judgment based on this mere message. Yet combining it with the tutor's observation of her own, it was confirmed that the learners had not acquired the functions of the CAT tools fully.

It was even worse when it came to the learners' ability to evaluate the CAT tools of their choice. As already shown in some learners' own reflections, they suggested cancelling this part of tool evaluation altogether as they could not see why they, as amateurs, had to do it. The evaluative report of the adopted CAT tools was without exception the worst part in the final assessment task. The groups just made a list of the problems they met with the tools during the project followed by some very general comments on the layout or certain specific functions of the tools. No systematic review was found of the functioning of the tools in light of the purposes they were intended to serve. This showed to the author that all the learners failed to achieve ILO 8 satisfactorily.

Lastly, regardless of the expressed zeal for group work, the observer noticed some groups did not work as efficiently as others, due to different factors, such as the personality of the group leader, the evaluation system adopted within the group, and the way of the work division. Again the students had done a good job to have been able to reach such a level within such a short time, but surely there is still great room for improvement in terms of their group work skills.

6.5   A summary

In this chapter, the triangulated data of both self-reported and factual kinds were examined and analysed in search of the results of the preliminary test of the PBL approach to CAT teaching at the course level. A common set of codes in alignment with the ILOs (see Table 3. 4 in Section 3. 3. 3. 3. 5) were adopted to guide the theory-driven data reduction process. The findings were reported above within the 3-level evaluation framework (Kirkpatrick 1998) with an aim to provide systematic evidence for the effects of the approach. Self-reported outcomes were further verified by the factual data from the tutor and the observer, to make the final conclusions more reliable.

To sum up, the analysis of the data produced by the case study shows us on the one hand highly positive effects of the preliminary test of the PBL approach designed for the CAT course, and, on the other, the suggestions for and concerns with the newly adopted approach and the ILOs it failed to achieve which point to us the future directions to pursue.

Firstly, at the level of reaction, all the participants expressed great satisfaction with the adopted approach, with some even highly grateful for the tutor's introducing them to a new world of translation by this fresh learning approach. Nevertheless, concerns and suggestions were also gathered for improving further implementation of this approach in broader contexts. They covered a wide range of issues, including scaffolding, learner characteristics, the tutor's role, class sizes, grouping methods, and problem design.

Secondly at the level of learning, the observed acquisition of knowledge and skills as well as positive attitude were quite obvious, although part of the advanced functions of SDL Trados 2011 seemed to be too difficult and some steps in the project management too complex to master within such a short span of the course.

Lastly changes in the learners' behaviour emerged in both the participants' recording, the tutor journals, the observational sheets as well as the assessment documents. While obvious progress was noticed in the learners' ability to learn independently, skills to cooperate with (and tolerate) group members, broad-sense CAT and narrow-sense CAT application skills, and the ability to apply acquired knowledge in the translation projects, some abilities addressed as

ILOs of this course were still found undesirably poor. Among them are less satisfactory ability to identify their own problems and think independently, failure to grasp the functions of the CAT tools fully, poor ability to evaluate CAT tools and better cooperative skills in group work to be expected.

Chapter Seven
Discussion and Conclusions

In previous chapters of this thesis, the author have reported in turn how this educational design research exploring a PBL approach to the CAT course was designed and field tested with results presented in Chapter Six. This chapter then intends to foreground the findings against the research questions. Pedagogical implications will be drawn for the CAT teaching in translation programmes and directions for future studies will be highlighted based on the new knowledge generated from the current study. Lastly limitations of this research will be pointed out before the conclusions are made.

7.1 The research questions revisited

The main purpose of this research was to introduce PBL into CAT teaching in college translation programmes. It attempted to justify the alignment between them in theory (i.e. research question 1) before piloting a design of PBL to the CAT course (i.e. research question 2) and field testing it as a case study in SIS, SYSU (i.e. research question 3). Chapters Four and Five have presented the theoretical connections and the pilot design of the PBL approach to the CAT course for SIS, SYSU. In this chapter, the author intends to discuss the findings of the preliminary test, which were reported in Chapter Six, in light of the third research question which integrates the objectives of 3 and 4.

As shown in the previous chapter, the results of the preliminary field test of the PBL design were quite encouraging. The data of both kinds, the self-reported and the factual, show almost consistently the positive outcomes of the CAT course at all the 3 levels of reaction, learning and behaviour (Kirkpatrick, 1998), which were allegedly attributable to the PBL approach adopted. The huge amount of time the participants showed to have invested in the course did convince us of their positive attitude and great devotion to this learning experience with the PBL approach. Their responses to Question 9 of the survey delivered one year after the course was over provided us with further evidence for the changes of the learners' behaviour resulting from their learning in the course. Therefore, the findings of this study are largely consistent with many previous qualitative studies on student perception of PBL (e.g. Taylor, 1997; Savin-Baden, 2000a; Wilkie, 2002 cited from Savin-Baden, 2003).

7.1.1　PBL and attainment of the ILOs

This section will look back at the ILOs the tutor set up for the course using the PBL approach and see how much they have been achieved.

As reported in the previous chapter, the reflective reports, journals and group study records, as well as the questionnaire survey were examined to find out evidence for the leaners' acquisition of required knowledge and skills. The self-reported data were also triangulated with those from the tutor journal and the final summative assessments for improved reliability. The results showed to our delight that all the ILOs had been attained although to different degrees. Direct quotations of the learning outcomes from the participants' report included not only the declarative and functional knowledge about "broad-sense CAT tools" (cf. ILOs 1, 2, 4, 5), "narrow-sense CAT tools" (i.e. SDL Trados 2011, Oriental Yaxin, Google Translator Toolkit) (cf. ILOs 1, 2, 3), but also such desired qualities of independent learners (cf. ILO 12) as "stronger motivation", "better ability to study independently", "better transferability of the learned knowledge and skills", "more reflective", and "more inclined to question before receiving the answers" and of teamplayers (cf. ILO 10) as "better group work skills" and "more willingness to share". Factual data, especially the assessment documents further verified that the learners did not only know about the tools but also know how to apply them to real projects. The acquisition of the functional knowledge among the ILOs, namely ILOs 6, 7, 8, 9, and 11, was clearly demonstrated in the fact that they all went through satisfactorily a simulate translation project (covering planning, implementing, evaluating and reflecting) using CAT technology.

Without a control group to compare with, no definitive conclusion could be made here to attribute all the learning outcomes to the adoption of the PBL approach. Indeed when comparing the participants' performance in the last assessment with that of students enrolled in the CAT course in the previous normal term (who were tested in the same way), the author did not see noticeable superiority of the learners in this study to the best performers then. Yet the statistical comparison of the means of the final scores of the whole classes did show some difference. The mean of the final scores in the normal term was 88 while the mean in this study was 93. 6. This difference is still subject to explanation based on further studies, but one of the possible reasons could be found nevertheless among the participants' own comments. Of the leaners' reflections, the learning outcomes most visibly attributed to the PBL approach were their enhanced motivation and willingness as well as the improved ability to learn independently. These in turn make it possible that the learners studied more proactively and invested more time in this course which would almost naturally result in the better learning effects.

Yet as commented by some leaners, although evidence existed for the learners' increased declarative knowledge of CAT technology, without a formal, close-book exam, it is hardly clear about how much of such knowledge had been acquired. But according the few students showing concerns in this regard, probably less content knowledge was obtained than in the normal term where the tutor would normally give two to three lectures introducing systematically the development and the working principles of CAT technology as well as their applications. However, the students at their current level could hardly have sorted out the information to the same effect.

Besides, the advantage of the PBL approach in arousing

students' interest to continue studying CAT is fairly obvious compared to the traditional approach according to the author's teaching experience. The author was further assured of the power of the PBL approach in this aspect when told by Zhang Zixi (张紫曦), a self-claimed nastiest technology learner ever, that she had successfully overcome the long persisting fear of technology and was interested to know more about CAT and when another student, Hong Shi (洪奭), who, convinced of the authoritativeness of teachers, had been long repelled by the idea of self-directed learning, expressed to the author that she wanted to re-enroll in my course in the long term to learn more about CAT and meantime to experience PBL once more.

In sum, the CAT course with the PBL approach has successfully enabled the participants to achieve the ILOs, with average performance better than the learners taught in the traditional approach based on the comparison between the mean score of the whole class in the final assessment of the translation project. The qualitative analysis of the effects of the course perceived by the learners shows us the remarkable power of PBL in enhancing learners' motivation, independent learning ability and group work skills. While most of the learners claimed to have acquired the required declarative and functional knowledge about CAT which was also largely supported by the factual data collected via tutor journals and classroom observation, due to the limitations of the assessment forms, the exact effect of the PBL approach on the declarative knowledge acquisition remains unknown. Yet the few learners' concern with the possibly less content knowledge acquired from the course was found in many other studies (Angeli, 2002; Lieux, 2001; Schultz-Ross & Kline, 1999, cited from Hung et al., 2008) and therefore accepted as a viable issue to be further explored later.

7.1.2 *Pedagogical implications*

All in all the very first try of the PBL approach in the CAT course gives us greater promise than doubt. The overall positive reaction toward it and the articulated learning outcomes, with some even out of expectations of the learners themselves, all allow us to report the following successful experience in this pedagogical renovation with the hope that teachers of CAT courses in the similar context may benefit from it.

7. 1. 2. 1   Collaborative learning in small groups

Among all the comments on the way of learning designed in the PBL approach, study by groups won the most acclaim. What the participants reported to have benefited from it are as follows.

Firstly, learners could learn from each other what they can never get from the tutor. With the same standing and similar background, learners would more easily or even more willingly perceive the strong points of their peers.

Secondly, peer pressure helped to push them beyond their limits. Working actively under the pressure from the task or simply from their more capable peers (as shown in [Quote 7] and [Quote 9]), the learners were compelled to step out of their "comfortable zone" and tap their potentials that they themselves might not be aware of. Such new findings of their potentials were even more exciting and encouraging for the learners to pursue further studies actively than simple acquisition of knowledge from others.

Thirdly, in group study, the students work toward a common goal together. Such awareness of somebody else erring and progressing at the same time makes learners feel less lonely and consequently stronger and more motivated to take efforts. Besides, after witnessing the pains their peers had gone through, the learners in their reflections on their experience in the course did much less complaining than self-criticism and appreciation of the superior qualities of their peers, forming then a right attitude towards learning.

Fifthly, diversity in groups creates a better environment to foster problem solving ability. Different learners bring in group study different perspectives which could help expand individual's horizon. Being aware of the possibility for different solutions to a problem, leaners will then become more flexible and more resourceful when dealing with problems themselves.

Lastly, the learners have polished their skills of communication and cooperation with peers via group study. The journals and the reflective reports showed us that they had realized that a sense of responsibility and reasonable division of work are key to the successful team work. Selecting members of different specialties, establishment of a common goal, well-thought planning and mutual understanding when in difficulty were all mentioned as experience they had acquired through group study in this course.

All the above gains make group study highly recommendable for not only CAT courses but also translation courses in general, as it serves very effectively for the goal of fostering professional translators who have self-concept (Kiraly, 1995) and can fulfil tasks in accordance with social conventions (Colina, 2009).

7. 1. 2. 2　Problem-driven tutorial process

Although there were not many comments directly addressing the tutorial process starting with problems, it is still worth recommending seeing the highly encouraging learning outcomes of this course.

As the quotes from the learners reflections in Section 6. 2. 3 show, when confronted with problems without any prior teaching, many students felt unaccustomed and even lost during the first a few days. That is probably why when making suggestions on what to improve in the course, some of them mentioned increasing tutor guidance, especially in the initial stage, and some proposed the addition of the traditional teaching method, namely lectures, to it.

Nevertheless, many more of the participants recognized their harvest from the PBL approach which is different, if not better, than that from the traditional teaching method. For example, one student described the feeling of finally overcoming all the difficulties via her own explorative study as extremely wonderful like "seeing sunshine after days of raining".

More specifically, after experiencing such a learning process pushed forward by problems instead of teachers, most of the participants, when describing their changes in the following year, mentioned their greater sensitiveness to problems, the growing inclination to ask their own questions and to try to find their own solutions to the problem before seeking for help from others, and in consequence better assimilation of the newly acquired knowledge.

It is widely acknowledged that translation is a process of problem-solving (Reiß, 2000; Sharoff, 2006). To cater to the new demands of modern society for professional translators, it has been long advocated that TTPs should set as their goal developing students' ability to "employ available knowledge to solve new problems and to gain new knowledge as the need arises" (Bernardini, 2004, p. 20). Therefore, the PBL tutorial process started by problems is similar to what is really happening in the professional world of translation and, no wonder, was found to produce better transferability of the acquired skills among the learners (see quotes in Section 6. 2. 3. 2).

7. 1. 2. 3   Self-directed learning

According to the findings of this study, the self-directed learning component of the PBL approach seemed to have at least the following strengths:

Many students expressed less reliance on teachers and classes after getting used to managing their own learning in the course. As the responses to Question 9 in the survey show, such embracement of independent learning enabled them not only to acquire the knowledge and skills in the course but to retain them much longer. Evidence for this is that when asked about whether they still remembered the content of the course they took one year earlier, they were not merely noting down what they had learnt; rather two of them recalled in great detail how they had been made to learn it. Better still, some students mentioned (see quotes in Section 6. 2. 3. 2) the habits they had grown in the course have been consciously or unconsciously transferred to their other activities in life. Among them the most obvious one is seeking actively for tools and methods to solve problems on one's own.

Moreover, as the power to decide what and how to learn was passed down to the students with the tutor retreating to the background only responsive to students' needs, some of the participants used "democratic" to describe their feeling about the process. The by-products thereof are the increased flexibility of the course in addressing different needs of different individuals as the students were allowed to proceed at their own pace and the enhanced confidence of the learners in their own ability to learn and relearn after they successfully conquered the problems in the course which had seemed almost impossible for them, as well as " deeper understanding" and "more critical and logical thinking".

All these findings suggest to the author the great power of such a self-directed learning process in fostering in learners the ability to identify, analyzing and solving problems independently and its high worth to be applied to broader contexts in higher education, as independent learning is a highly desirable quality of graduates from translation programmes and beyond in the 21st century (Kaiser-Cooke, 1994; Wilss, 1996; Bilić, 2013).

7. 1. 2. 4　Context of PBL implementation

The pilot design of the PBL approach was field tested in the credit-bearing CAT course project in SIS, SYSU. The context of the preliminary implementation in this case study will be reiterated below so that the experience of the current study can be of some referential value to CAT course designers elsewhere.

First of all, SYSU is among top ten universities in China. The CAT course has been on the curriculum of the Direction of Translation and Interpreting, one of the four directions[①] within English major in SIS, ever since the year of 2005 when it was founded. The administration of SIS has shown great support for this course, equipping it with a computer lab and two mainstream commercial CAT systems, namely SDL Trados and Oriental Yaxin.

Secondly, the students enrolled in the course are junior English majors in Department of Translation and Interpreting who had never taken any translation courses and who had long been used to the traditional transmissionist way of teaching.

[①]　The four directions form administratively four separate departments in SIS, namely Department of Translation and Interpreting, Department of Business Communications, Department of Chinese as a Foreign Language, and Department of International Affairs.

Thirdly, the study was conducted during the third term which was much shorter than the spring and autumn terms and set aside for more specialized training of students in light of the special needs of different disciplines. Courses designed for the short term in SIS were therefore comparatively more professionally oriented and students were spared more free time to develop their own interests. So far as the author was concerned, the majority of the participants had selected only this course during the four weeks. That is probably why the participants of the study could afford so much attention to the course as shown in the amount of time they had invested in it.

Fourthly, the participants were selected with a set of criteria to strike a balance between students of different characteristics and limited in the total number under 25. All this might be hard, if not impossible, to realize in the normal terms. Besides, the participants happened to be highly motivated and cooperative, which made this innovative attempt much easier.

Lastly, the tutor herself could also devote more time to the course as she was teaching only this course and free from other administrative affairs which is usually a big burden in normal terms. Besides, she had had nearly 8 years of teaching CAT in SIS and studied PBL for a while when she offered this course with the PBL approach.

To sum up, in consideration of a shift to the PBL approach, the factors that have to be taken into account are the school it is applied in and the attitude of its administration toward such innovation, the characteristics of students, the tutor's experience, the class size, and the availability of sufficient time and energy of both the participating students and the tutor.

7.1.3   Directions for future studies

As summarized in the previous chapter, both the self-reported data and the factual data have revealed to us suggestions for improvements and issues worth further exploration. They combine to point to us the directions for the future studies which will be specified below.

7. 1. 3. 1   Effective scaffolding

In this study the author adopted Walqui's (2006) framework of scaffolding as three related pedagogical scales (see Section 5. 1. 2. 3), which, in this case, include the planned teaching progression (i.e. the problems), the tutorial process, and the collaborative process of interaction. The former two provide students with a supportive structure, which is relatively stable, and the last one collaborative construction process which is more improvised, contingent on learners' needs.

As reported in Chapter Six, among the suggestions for improvements the learners were invited to make, those concerning scaffolding are the most. Among them are more explanation about what to learn and how to learn with the PBL approach, more guidance on how to learn independently, a few lectures to inform learners of the key content, and an additional timetable common to the class for when to finish what readings.

These suggestions, as analysed before, show to the author, for one thing, the difficulty the learners experienced especially at the initial stage of the course in transiting from the traditional learning approach to PBL. Students' transition from traditional methods to PBL has attracted much attention from researchers for some time. For example, Schmidt, Boshuizen & de Vries (1992) found that it took at least 6 months for students to adapt to the new approach. Many more researchers were interested in what caused this difficulty and among the suggestions are uncertainty about their grades (Woods, 1994, 1996) and difficulty in taking more responsibilities in the learning process by learners stuck in their roles in the traditional method (Dean, 1999; Lieux, 2001). Therefore, scaffolding in early stages of the course constitutes an important issue to explore to find out how to facilitate learners' transition to the new approach so as to benefit more from it.

For the other, these suggestions remind the author of the learners' failure to grasp fully the functions of the required CAT tools and their falsely disregarded value of the evaluation tasks in the translation project. These less satisfying learning outcomes indicate to the author, on the one hand, the leaners' independent learning competence to be further enhanced, and, more importantly, the possible weaknesses in the scaffolding design on the other.

Of the three scales of scaffolding, all the suggestions from the learners involve only the third one. What is surprising is that although many students were eager to get more of the tutor guidance, Group 5 expressed a different opinion (see Section 6. 2. 2). They found themselves easily influenced by the tutor during interaction and preferred to be allowed for more space to work on their own. This contrast in learners' suggestions introduces a new factor into consideration when thinking of improving the interactive and collaborative scaffolding process, that is, student need.

Without prior experience in learning with PBL approach, the learners were not capable of noticing the problems with the structure scaffolding, namely the problem sequence and the tutorial process. Yet as reviewed above, the less desirable outcomes of leaners pointed out to us there is big room for improvement.

In a nutshell, well-designed scaffolding is key to successful implementation of the PBL approach, especially in similar context to the current one where learners experience PBL for the first time. Measures to take to facilitate the initial stage of transition, proper structuring of the problems and the tutorial process corresponding to the ZPD of learners, and timely collaborative interaction offered by the tutor rightly responsive to students' needs are all the fields worth further studies in order for the PBL approach to produce expected outcomes.

7. 1. 3. 2 Tutor roles

Barrow (1992) defined the tutor's roles with two responsibilities: one is facilitating students' acquisition of the desired knowledge and skills and the other is helping them become independent and self-directed learners. The new roles were attached great importance to (Maudsley, 1999) for the effectiveness of PBL and naturally posed a great challenge to educators who have to redefine the nature of learning and reposition themselves from the traditional knowledge transmitter to a learning process facilitator.

Therefore, research has found the tensions arising in teachers experiencing such a paradigm shift and their discomfort about the new roles as an almost inevitable result. The author felt the same way regardless of her heart-felt embracement of the learning philosophy underlying PBL and her preparation both intellectually and psychologically for this shift. In her journals, she noted many times that she found it very hard to resist the desire or merely to drop the habit of talking too much. To look back at most of the time she could not help but control the interaction with the learners, we found that it happened either because she wanted to keep the conversation going but got scarce responses from the learners who happened to have a knowledge gap in the topic under discussion or because she just fell back to the traditional role unconsciously. Evidence of this kind shows us that the tutor also will undergo a hard time experiencing difficulty in redefining their roles and practicing it in real teaching.

What makes the situation even more complex is the diversity in students' characteristics and their very individualized needs. As the quotes in Section 6. 2. 3 show, students may seek anxiously for help from the tutor when stuck in trouble yet may as well feel disappointed at themselves if they could not contribute much to the solution to the problem. Such a dilemma for tutors of PBL has begot many studies on the tutors' roles and multiple roles of the tutor have been proposed, such as information disseminator, evaluator, parent, professional consultant, confidant, learner, and mediator (Wilkerson & Hundert, 1991).

As recognized worldwide, Chinese learners are more passive than students of the west. Then to further implement PBL in CAT teaching in China, what the tutor would have to do to strike a balance between her two responsibilities deserves more attention from our native researchers.

7. 1. 3. 3　Problem design

At the centre of the PBL approach, problems initiate and organize the whole learning process and, no doubt, are crucially important for the success of PBL (Duch, 2001; Trafton & Midgett,

2001, cited from Hung *et al.*, 2008).

Seeing the paramount importance of it, the fact that few students actually made suggestions on the problems design does not relieve the author from the concern with their quality and effectiveness. The four problems in this course were designed following the four principles specified in Section 5. 1. 2. 2. 3 and were aligned with the ILOs. Yet due to the limited scope of the current study, how exactly the problems have abided by the principles and how able they were to stimulate and organize learning towards the ILOs were not able to be examined in depth. The few suggestions concerning problem design reflected more of the weaknesses in scaffolding provided than with the design of the problems.

According to Hung *et al.* (2008), problem design is still less investigated than necessary and it is no exception in the context of CAT teaching. In order to understand better how the problem can be designed to effectively enable learning towards the ILOs, the degrees of complexity and structuredness of CAT problems usually encountered by the professionals, the model and principles of problem design and effectiveness of the problems in PBL are inarguably three most urgent issues for future research.

7. 1. 3. 4   Assessment design

As is widely recognized, it is the assessment rather than the ILOs that defines students' learning, whether we like it or not. Assessment, as a result, is highly valued by PBL designers to achieve maximum positive backwash on students. Yet as Savin-Baden (2004) pointed out, assessment is probably among the most controversial issues in PBL as it is most often used to validate the effectiveness of PBL.

As research in this regard increases, people have begun to realize that the reliance on the traditional standardized tests has misled them in their judgment about the effectiveness of PBL. It is based on the experience that the author adopted the framework of CA (Biggs, 2007), try to keep what the learners learn and what they are tested about consistent. No complaint from the participants is an affirmative signal to the author. Yet opportunities for more research in this regard abound. First of all, to fully motivate the learners, the author integrated in her assessment design both process-oriented and outcome-oriented forms, both tutor-controlled and peer-as well as self-evaluation, and both group-based and individually based grading methods, which made the grading procedures rather complicated and the tutor much too burdened. Moreover how each of the aspects can be weighted in a reasonable and valid way is another big problem. Therefore, further studies are expected to find a way to strike a balance between the complexity and the efficiency of the assessment design. Next, no less important is better-controlled research on the validity and reliability of the translation project as the assessment task to address the ILOs. Besides, the exact effects remain unclear of the different components in the translation project assessment on testing content knowledge and applied knowledge.

7. 1. 3. 5　Student characteristics

The issue of student characteristics has actually drawn the researcher's attention when selecting the case for the field test of the pilot design. Four characteristics of students, namely gender, academic performance, prior experience and their shown interest in pursuing translation as their future career, were considered in selecting the participants and in grouping them. Although few comments were found on the influence of these four characteristics on their learning effects, when invited for suggestions, the participants did express their hope for more attention paid to the diversity of students' strengths when doing the grouping (see Section 6. 2. 3). During some in-class conversations with the students, the tutor once asked about the function of gender in their learning of CAT. The answers were almost unanimously "no noticeable effect". Rather *computer literacy* seems to be a more desirable characteristic irrespective of gender. Moreover, academic performance was not found to be interfering with students' learning of the technological part of the course although it was indeed quite important during the translating phase as observed by the tutor. Therefore it remains unclear yet important to know what characteristics of learners would influence the effect of learning with the PBL approach, which does not only decide the learner's gains from it but also affects how well the group study can actualize its potentiality.

7.2   Limitations of this study

At the initial stage of the PBL design research, the current study was aimed to explore the possibility for application of PBL to CAT teaching in higher education by experiencing and understanding the process. Qualitative methodology was employed given its congruency with the purpose of this study on the one hand and its consistency with the underlying philosophy of PBL on the other. Nevertheless qualitative research has limitations in itself. First of all it is most often criticized for lacking scientific rigour in terms of mainly objectivity and generalizability (Berg, 2007).

Firstly in terms of the objectivity, the author admits that relying the evaluation of the course effects largely on the participants' self-report data reduced its trustworthiness. The theory-driven thematic analysis of the data might as well miss out some information. Moreover the author as the only researcher to interpret such data is likely to distort the data with personal bias which would have resulted in somewhat uncertainty in the reproducibility of the study. That is, a different researcher may not come to the same interpretation as it is now.

Secondly the design and implementation were all carried out in the context of SIS, SYSU. The case study focuses on a small number of research settings and therefore almost inevitably brings challenges for its generalizability (ibid.).

The author was highly aware of these built-in shortcomings of the research design, so she first of all made clear the compatibility of case study research and the current study in exploratory design stage and set for her research objectives manageable within the limits of the methodology adopted. Meanwhile, she took some measures to guarantee a reasonable degree of validity of the research findings. Among them the most effective one should be the triangulation of data sources and data collection instruments, namely data and methodology triangulation (Guion *et al.*, 2011). Besides, the author articulated the research procedures in detail and revealed her own beliefs and stands concerning PBL. By making the study process transparent, it is hoped that the subjectivity of the research is reduced and the replication enhanced simultaneously.

Being still in the exploratory stage of the PBL design research, this study was aimed to build theoretical connections and attempted at understanding how the pilot design worked on learners by preliminary teaching in a case study. Although there did exist some shortcomings in the research design, the research questions put forward at the beginning of the research have all been answered at least satisfactorily. Generally speaking, the applicability of PBL to CAT teaching is affirmed at course level in college translation programmes as it has been received very positively by the participating students and is believed to have turned out all the ILOs of this course with most obvious effects on enhancing students' motivation to learn CAT and their independent learning and teamwork cooperation abilities. More focused positivist research with better control is desired to improve the design and further explore the effectiveness of the different components respectively in the PBL approach to CAT teaching.

7.3 Conclusions

The aim of this study was to address the challenges faced with the CAT teaching in higher education, resulting from the changes in the professional world and the advancement of the learning theories. In view of the undesired neglect of CAT teaching methodology among researchers, the author made an attempt at introducing PBL into a CAT course at college level via an educational design research project at SYSU. The adoption of PBL was inspired by the new approach's widely reported promising power, especially in professional fields like science, social studies, business and medicine, of developing students' domain-specific knowledge and skills as well as their generic skills such as self-directed learning, collaborative work, and problem-solving that are highly expected of college graduates in the 21st century.

With this three-stage design research, theoretical connections were first built up by examining the compatibility between PBL and the combined demands from the professional world and the higher education. The systematic exploration of the PBL approach to CAT course design was then accomplished after the concept clarification and framework establishment. Following it the pilot design was field tested in a case study in SIS of SYSU by which preliminary evidence was successfully gathered for applicability of the PBL approach to CAT teaching in higher education. Data analysis shows us that the ILOs set for the course were found to be largely achieved and the learners' motivation, self-directed learning, group work collaboration and problem-solving skills were greatly enhanced as evidenced by both the learners' self-report perception and the factual data from the assessment tasks and the observational sheets. The triangulated data collection method assures us of the validity of the data. Yet the limitations of the research methodology and the scope of the current study leave many issues to be explored by research with better control in future in order to further improve the PBL design and ensure its application to a wider context, such as scaffolding, tutor roles, problem design and student characteristics.

It was our intention for the time being to examine the PBL process with only this particular group of students in SIS, SYSU, the findings of which were therefore not intended to be generalizable to other universities. Yet our thoughts and experience may well be informative to other CAT teachers considering a pedagogical renovation or curious about the effects of the PBL approach. It is hoped that this study has not only made contribution to the pedagogical design research in CAT teaching studies but also provided more empirical support for the applicability and validity of PBL in translation education at tertiary level.

References

Akker, V. J., Gravemeijer, K., McKenney, S. & Nieveen, N. (2006). *Educational Design Research*. London: Routledge.

Alcina, A. (2008). Translation technologies: Scope, tools and resources. *Target*, 1, 80-103.

Alcina, A., Soler, V. & Joaquín, G. (2007). Translation technology skills acquisition. *Perspectives*, 15 (4), 230-244.

Alfieri, L., Brooks, J. P., Aldrich, J. M. & Tenenbaum, R. H. (2011). Does discovery-based instruction enhance learning? *Journal of Educational Psychology*, 103 (1), 1-18.

Allen, D. E. & White, H. B. (2001). Undergraduate group facilitators to meet the challenges of multiple classroom groups. In J. B. Duch, E. S. Groh & E. D. Allen (2001a, pp. 79-94).

ALPAC (Automatic Language Processing Advisory Committee) (1966). *Language and Machines: Computers in Translation and Linguistics*. Washington, D. C. : National Academy of Sciences National Research Council.

Anderson, L. W. & Krathwohl, D. R. (eds.) (2001). *A Taxonomy for Learning, Teaching, and Assessing: A Revision of Bloom's Taxonomy of Educational Objectives*. New York: Longman.

Archer, J. (2002). Internationalisation, technology and translation. *Perspectives: Studies in Translatology*, 10 (2), 87-117.

Arnold, D., Balkan, L., Humphreys, R. L., Meijer, S. & Sadler, L. (1994). *Machine Translation: An Introductory Guide*. Oxford: NCC Blackwell.

Austermühl, F. (2001). *Electronic Tools for Translators*. Manchester: St. Jerome.

Baer, B. J. & Koby, G. S. (2003). Introduction: Translation pedagogy: The other theory. In B. J. Baer & G. S. Koby (eds.) *Beyond the Ivory Tower* (pp. i-viii). Amsterdam/Philadelphia:

John Benjamins.

Balkan, L. (1992). Translation tools. *Meta: Translators' Journal*, 37 (3), 408-20.

Barab, S. & Squire, K. (2004). Design-based research: Putting a stake in the ground. *The Journal of the Learning Sciences*, 13 (1), 1-14.

Barak, M. & Dori, Y. J. (2005). Enhancing undergraduate students' chemistry understanding through project-based learning in an IT environment. *Science Education*, 89 (1), 117-139.

Bar-Hillel, Y. (1960). The present status of automatic translation of languages. *Advances in Computers*, (1), 91-163.

Barrett, T. (2005). Understanding problem-based learning. In Barrett, T., Mac Labhrainn, L. & Fallon, Hco. (eds.). *Galway: CELT* (pp. 13-25). Retrieved on June 26, 2014 from http://www. nuigalway. ie/celt/pblbook/chapterl. pdf.

Barrows, H. S. (1986). A taxonomy of problem-based learning methods. *Medical Education*, (20), 481-486.

Barrows, H. S. (1992). *The Tutorial Process*. Springfield, IL: Southern Illinois University School of Medicine.

Barrows, H. S. (1996). Problem-based learning in medicine and beyond: A brief overview. *New Directions for Teaching and Learning*, (68), 3-12.

Barrows, H. S. & Kelson, A. (1995). *Problem-based Learning: A Total Approach to Education* (Monograph Series). Springfield, IL: Southern Illinois University School of Medicine.

Bazeley, P. (2007). *Qualitative Data Analysis with Nvivo*. London: Sage.

Berg, B. (2007). *Qualitative Research Methods for Social Sciences*. USA: Pearson.

Bernardini, S. (2004). The theory behind the practice: Translator training or translator education? In Melmkjær, K. (ed.), *Translation in Undergraduate Degree Programmes* (pp. 17-30). Amsterdam/Philadelphia: John Benjamins.

Biggs, J. (1996). Enhancing teaching through constructive alignment. *Higher Education*, (32), 347-364.

Biggs, J. & Tang, C. (2007). *Teaching for Quality Learning at*

University. Maidenhead: Open University Press/McGraw Hill.

Bilić, V. (2013). PBL meets translation: Emigrant letters as case studies. *T21N— Translation in Transition*, (5), 1-22.

Bolger, N., Davis, A. & Rafaeli, E. (2003). Diary methods: Capturing live as it is lived. *Annual Review of Psychology*, (54), 579-616.

Borg, W. R. & Gall, M. D. (1989). *Educational Research*. New York: Longman.

Boud, D. & Feletti, G. (eds.) (1991). *The Challenge of Problem-Based Learning*. London: Kogan Page.

Bowker, L. (2002). *Computer-aided Translation Technology: A Practical Introduction*. Ottawa: University of Ottawa Press.

Bowker, L. & Marshman, E. (2009). Better integration for better preparation: Bringing terminology and technology more fully. *Terminology*, 15 (1), 60-87.

Bridges, E. M. & Hallinger, H. (1995). *Implementing Problem-based Learning in Leadership Development*. Eugene: University of Oregon, Educational Resources Information Center, Clearinghouse on Educational Management.

Brooks, J. G. (2002). *Schooling for Life: Reclaiming the Essence of Learning*. Alexandria, VA: Association for Supervision and Curriculum Development (ASCD).

Brown, D. (1994). *Teaching by Principles. An Interactive Approach to Language Pedagogy*. New Jersey: Prentice Hall Regents.

Bruner, J. (1983). *Child's Talk: Learning to Use Language*. New York: Norton.

Campbell, J. R. (1999). *The Development and Validation of an Instructional Design Model for Creating Problem Based Learning*. Unpublished Doctor of Education Thesis, University of Pittsburgh, Pittsburgh, PA, USA.

Chai, M. J. (柴明颎) (2010). Dui zhuanye fanyi jiaoxue jiangou de sikao: Xianzhuang, wenti he duice (对专业翻译教学建构的思考：现状、问题和对策，"On construction of specialised translator education: The status quo, problems and solutions"). *Zhongguo Fanyi* (中国翻译，"Chinese Translators Journal"), (1), 54-56.

Chan, Sin-wai (ed.) (2007a). *Readings in Computer-Aided Translation* [*For TRA7008 (2007-08)* "*Introduction to Computer-Aided Translation*"]. Unpublished internal readings of CAT programme in the Chinese University of Hong Kong.

Chan, Sin-wai. (2007b). Computer-aided translation: Where do we go from here. In Chan, Sin-wai (2007a, pp. 3-7).

Charlin, B., Mann K. & Hansen P. (1998). The many faces of problem-based learning: A framework for understanding and comparison. *Medical Teacher*, 20 (4): 323-330.

Charmaz, K. (2001). *Constructing Grounded Theory: A Practical Guide Through Qualitative Analysis*. London: Sage Publishers.

Chellamuthu, K. C. (2002). Russian to Tamil Machine Translation System at TAMIL University. In *Proceedings of INFITT 2002, California, U. S. A.* (pp. 74-83). Retrieved June 20, 2012 from http://www. infitt. org/ti2002/papers/16CHELLA. PDF.

Chen, C. W., Feng, R. F. & Chiou, A. F. (2009). Vygotsky's perspective applied to problem-based learning in nursing education. *Fu-Jen Journal of Medicine*, 7 (3), 141-147.

Chen, N. C. (2008). An educational approach to problem-based learning. *Kaohsiung Journal of Medical Sciences*, 24 (No. 3 Suppl): S23-30.

Clark, R. (1994). Computer-aided translation: The state of the art. In C. Dollerup & A. Lindegaard (eds.), *Teaching Translation and Interpreting 2: Insights, Aims, Visions* (pp. 301-308). Amsterdam/ Philadelphia: John Benjamins.

Clayton, A. & Thorne, T. (2000). Diary data enhancing rigour: Analysis framework and verification tool. *Journal of Advanced Nursing*, (32), 1514-1521.

Cohen, S. A. (1987). Instructional alignment: Searching for a magic bullet. *Educational Researcher*, 16 (8), 16-20.

Colina, S. (2009). *Translation Teaching from Research to the Classroom: A Handbook for Teachers*. Shanghai: Shanghai Foreign Language Education Press.

Cronin, M. (2010). Deschooling translation: Beginning of century reflections on teaching translation and interpreting. In M. Tennent (ed.) (pp. 249-265).

Cui, Q. L. (崔启亮) (2012). Gaoxiao MTI fanyi yu bendihua kecheng jiaoxue shijian (高校 MTI 翻译与本地化课程教学实践, "On teaching translation and localization courses in MTI"). *Zhongguo Fanyi* (中国翻译, "China Translators Journal"), (1), 29-34.

Davies, M. G. (2010) Undergraduate and postgraduate translation degrees: Aims and expectations. In K. Malmkjær (ed.), *Translation in Undergraduate Degree Programmes* (pp. 67-82). Shanghai: Shanghai Foreign Language Education Press.

Davies, M. H. & Harden R. M (1999). AMEE Medical Education Guide No. 15: Problem-based learning: A practical guide. *Medical Teacher*, (21), 130-140.

Dean, C. D. (1999). Problem-based learning in teacher education. Paper presented at the Annual Meeting of American Educational Research Association, April 19-23, Montreal, Quebec (ERIC Document Reproduction Service No. ED 431-771).

Donnelly, R. (2004). Investigating the effectiveness of teaching "on-line learning" in a problem-based learning on-line environment. In M. Savin-Baden & K. Wilkie (eds.), *Challenging Research into Problem-based Learning* (pp. 50-64). Berkshire: Society for Research into Higher Education & Open University Press.

Duch, B. (2001). Writing problems for deeper understanding. In J. B. Duch, E. S. Groh & E. D. Allen (eds.) (2001a, pp. 47-53).

Duch, J. B., Groh, E. S. & Allen, E. D. (eds.) (2001a). *The Power of Problem-Based Learning: a Practical "How To" for Teaching Undergraduate Courses in Any Discipline*. Sterling, VA: Stylus.

Duch, J. B., Groh, E. S., & Allen, E. D. (2001b). Why problem-based learning? A case study of institutional change in undergraduate education. In J. B. Duch, E. S. Groh & E. D. Allen (eds.) (2001a, pp. 3-11).

Engel, C. E. (1991). Not just a method but a way of learning. In D. Boud & G. Feletti (eds.) (pp. 22-33).

EMT Expert Group (2009). Competences for professional translators, experts in multilingual and multimedia communication. Retrieved July 29, 2012 from http://ec. europa. eu/dgs/translation/programmes/emt/key_documents/emt_competences_translators_en. pdf .

EPPI-Centre (2010). *Methods for Conducting Systematic Reviews.* Retrieved February 26, 2014 from http://eppi. ioe. ac. uk/cms/ Default. aspx? tabid =88.

Ertmer, P. A., Glazewski, K. D., Jones, D., Ottenbreit-Leftwich, A., Goktas, Y., Collins, K., et al. (2009). Facilitating technology-enhanced problem-based learning (PBL) in the middle school classroom: An examination of how and why teachers adapt. *Journal of Interactive Learning Research,* 20 (1), 35-54.

Finkle, S. L. & Torp, L. L. (1995). *Introductory Documents.* (Available from the Center for Problem-Based Learning, Illinois Math and Science Academy, 1500 West Sullivan Road, Aurora, IL 60506-1000.)

Finucane, P. M., Johnson, S. M. & Prideaux, D. J. (1998). Problem-based learning: Its rationale and efficacy. *Medical Journal of Australia,* (168), 445-448.

Fosnot, C. T. (1996). Constructivism: A psychological theory of learning. In C. T. Fosnot (ed.) *Constructivism: Theory, Perspectives and Practice* (pp. 8-33). New York: Teachers College Press.

Freebody, R. P. (2003). *Qualitative Research in Education.* London: Sage.

Fulford, H. (2002). Freelance translators and machine translation: An investigation of perceptions, uptake, experience and training needs. *Proceedings of the 6th European Association for Machine Translation Workshop* (pp. 117-122). November 2002, UMIST, Manchester.

Fulford, H. & Granell-Zafra, J. (2005). Translation and technology: A study of UK freelance translators. *The Journal of Specialised Translation,* (4), 2-17.

Gabrian, B. (1986). The aim or aimlessness of translation teaching. *Text & Context,* 1 (1), 48-62.

Gall, D. M., Gall, P. J. & Borg, R. W. (2003). *Educational Research: An Introduction* (Seventh Edition). Boston: Pearson Education, Inc.

Gallagher, S. A. (1997). Problem-based learning: Where did it come from, what does it do, and where is it going? *Journal for the*

Education of the Gifted, 20 (4), 332-362.

Gerloff, P. (1988). *From French to English: A Look at the Translation Process in Students, Bilinguals, and Professional Translators*. Unpublished PhD Dissertation. Harvard University, Cambridge (MA).

Gijbels, D., Dochy, F., van den Bossche, P. & Segers, M. (2005). Effects of problem-based learning: A meta-analysis from the angle of assessment. *Review of Eductional Research*, 75 (1), 27-61.

Gil, B. & Pym, A. (2006). Technology and translation: A pedagogical overview. In A. Pym, A. Perekrestenko & B. Starink (eds.), *Translation Technology and Its Teaching (With Much Mention of Localization)*. Tarragona, Spain: Servei de Publications (pp. 5-10). Retrieved on July 10, 2012 from http://isg. urv. es/publicity/isg/publications/technology_2006/idex. htm.

Gile, D. (2011). *Basic Concepts and Models for Interpreter and Translator Training* (Revised Edition). Shanghai: Shanghai Foreign Language Education Press.

Glesne, C. (2006). *Becoming Qualitative Researchers: An Introduction*. New York: Pearson Publishers Inc.

Göpferich, S. (2009). Towards a model of translation competence and its acquisition: The longitudinal study *TransComp*. In S. Göpferich, A. L. Jakobbsen & I. M. Mees (eds.), *Behind the Mind: Methods, Models and Results in Translation Process Research* (pp. 11-37). Copenhagen: Samfundslitteratur Press.

Gouadec, D. (2007). *Translation as a Profession*. Amsterdam/Philadelphia: John Benjamins.

Graaff, E. & Kolmos, A. (2003). Characteristics of problem-based learning. *International Journal of Engineering Education*, 19 (5), 657-662.

Guion, L. A., Diehl, D. C. & McDonald, D. (2011). Triangulation: Establishing the validity of qualitative studies. Retrieved December 12, 2013 from http://edis. ifas. ufl. edu/pdffiles/fy/fy39400. pdf.

Hakkarainen, P. (2009). Designing and implementing a PBL course on educational digital video production: Lessons learned from a design-based research. *Educational Technology Research and*

Development, 57 (2): 211-228.

Harden, R. M. & Davis, M. H. (1998). The continuum of problem-based learning. *Medical Teacher,* 20 (4), 317-322.

Harland, T. (2003). Vygotsky's zone of proximal development and problem-based learning: Linking a theoretical concept with practice through action research. *Teaching in Higher Education,* 8 (2), 263-272.

Haynes, C. (1998). *Breaking Down the Language Barriers.* London: Aslib.

He, S. Q. (何少庆) (2010). Cong yizhe shengcun zhuangkuang diaocha baogao kan fuzhu fanyi gongju de yingyong (从译者生存状况调查报告看辅助翻译工具的应用, "Necessity of access to translation-assisting tools as inferred from the investigation report on tranlstors' living conditions"). *Zhejiang Shifan Daxue Xuebao Shehui Kexue Ban* [浙江师范大学学报 (社会科学版), "Journal of Zhejiang Normal University (Social Sciences)"], (2), 101-103.

Heather, F. & Joaquín, G. (2005). Translation and technology: A study of UK freelance translators. *The Journal of Specialised Translation,* (4), 2-17.

Hendry, G. D., Frommer, M. & Walker, R. A. (1999). Constructivism and problem-based learning. *Journal of Further and Higher Education,* (23), 359-371.

Hilal, H. A. Y. & Alabri, S. S. (2013). Using Nvivo for data analysis in qualitative research. *International Interdisciplinary Journal of Education,* 2 (2), 181-186.

Hmelo-Silver, C. E. (2004). Problem-based learning: What and how do students learn? *Educational Psychology Review,* 16 (3), 235-266.

Hmelo-Silver, C. E. & Barrows, H. S. (2006). Goals and strategies of a problem-based learning facilitator. *Interdisciplinary Journal of Problem-based Learning,* 1 (1), 21-39.

Hmelo-Silver, C. E., Duncan, R. G. & Chinn, C. A. (2007). Scaffolding and achievement in problem-based and inquiry learning: A response to Kirschner, Sweller, and Clark (2006). *Educational Psychologist,* 42 (2), 99-107.

Hoffman, B. & Ritchie, D. (1997). Using multimedia to overcome the problems with problem based learning. *Instructional Science*, (25), 97-115.

Hou, X. C. (侯晓琛) & Yan Y. C. (颜玉祚) (2008). Zhongguo fanyi hangye CAT yingyong xianzhuang diaocha baogao (中国翻译行业 CAT 应用现状调查报告, "A survey report on the uptake of CAT technology in translation profession of China"). Retrieved on November 25, 2011 from http://sinastorage. cn/fs/800/1/936cde47825846e447c100041690c3a925654956/rar/CAT% E5% 9C% A8% E4% B8% AD% E5% 9B% BD% E7% BF% BB% E8% AF% 91% E8% A1% 8C% E4% B8% 9A% E4% B8% AD% E7% 9A% 84% E7% 8E% B0% E7% 8A% B6% E8% B0% 83% E6% 9F% A5-% E6% 95% B4% E5% 90% 88% E7% 89% 884. rar? origin =dl22. d. iask. com

Huang, K. S. & Wang, T. P. (2012a). Applying problem-based learning (PBL) in university English translation classes. *The Journal of International Management Studies*, 7 (1), 121-127.

Huang, K. S. & Wang, T. P. (2012b). Utilizing problem-based learning (PBL) in a university english interpretation class. *The Journal of Human Resource and Adult Learning*, 8 (1), 7-15.

Hung, W., Jonassen, H. D. & Liu, R. D. (2008). Problem-based learning. In J. M. Spector, M. D. Merrill, J. Merriënboer & M. P. Driscoll (eds.) (pp. 485-506).

Hung, W. (2011). Theory to reality: A few issues in implementing problem-based learning. *Educational Technology Research and Development*, (59), 529-552.

Hutchins, W. J. (1998). The origins of the translator's workstation. *Machine Translation*, 13 (4), 287-307.

Hutchins, W. J. (2000). The IAMT certification initiative and defining translation system categories. In *Proceedings of 5ᵗʰ EAMT Workshop*, Slovenia. Retrieved March 2, 2012, from http://ourworld. compuserve. com/hompages/WJHutchins/IAMTcert. html.

Hutchins, W. J. (2005). The history of machine translation in a nutshell. Retrieved November 1, 2013, from http:// www. thocp. net/reference/machinetranslation/machinetranslation. html

Hutchins, W. J. & Somers, L. H. (1992). *An Introduction to Machine Translation.* London: Academic Press.

Inoue, I. (2005). PBL as a new pedagogical approach for translator education. *Meta: Translators' Journal,* 50 (4). Retrieved on January 27, 2014 from http://id. erudit. org/iderudit/019865ar.

Institute of Education Sciences, US Department of Education and the National Science Foundation (2013). *Common Guidelines for Education Research and Development.* Retrieved on December, 2012 from http://ies. ed. gov/pdf/CommonGuidelines. pdf.

Jääskeläinen, R. (1989). The role of reference material in professional vs. non-professional translation. In Tirkkonen-Condit, S. & Condit, S. (eds.) *Empirical Studies in Translation and Linguistics.* (Studies in Languages 17) (pp. 175-198). Joensuu: University of Joensuu.

Janesick, V. J. (1998). Journal writing as a qualitative research technique: History, issues, and reflections. Paper presented at the Annual Meeting of the American Educational Research Association. April 13-17, 1998, SanDiego, CA.

Jia, Q. Q. (2013). Jiangou yi fazhan fanyi nengli wei hexin de MTI biyi peiyang moshi (建构以发展翻译能力为核心的 MTI 笔译培养模式, "Constructing a translation-competence-based translator education model"). *Yingyu Jiaoshi* (英语教师 "English Teachers"), (3), 13-17, 22.

Jonassen, D. H. & Hung, W. (2008). All problems are not equal: Implications for PBL. *Interdisciplinary Journal of Problem-Based Learning,* 2 (2), 6-28.

Kaiser-Cooke, M. (1994). Translatorial expertise a cross-cultural phenomenon from an interdisciplinary perspective. In M. Snell-Horby, F. Pöchhacker & K. Kaindl (eds.), *Translation Studies: An Interdiscipline* (pp. 135-139). Amsterdam/Philadelphia: John Benjamins.

Kay, M. (1997). The proper place of men and machines in language translation. *Machine Translation,* 12 (1-2), 3-23.

Kenny, D. (1999). CAT tools in an academic environment: What are they good for? *Target,* (1), 65-82.

Kerkkä, K. (2009). Experiment in the application of problem-based learning to a translation course. *Käännösteoria, ammattikielet ja monikielisyys*. VAKKI: n julkaisut, N: o 36. Vaasa 2009, 216-227.

Kim, B. (2001). Social constructivism. In M. Orey (ed.), *Emerging Perspectives on Learning, Teaching, and Technology*. Retrieved June 26, 2014 from http://projects. coe. uga. edu/epltt/.

Kiraly, D. C. (1995). *Pathways to Translation. Pedagogy and Process*. Kent: The Kent State University Press.

Kiraly, D. C. (2000). *A Social Constructivist Approach to Translator Education: Empowerment from Theory to Practice*. Manchester: St. Jerome.

Kiraly, D. C. (2001). Towards a constructivist approach to translator education. *Quaderns. Revista de traducci*, (6), 50-53.

Kirikova, L. & Šveikauskas, V. (2007). *Problem Based Learning: Student/Tutor Handbook* (Revision 1). Retrieved July 2, 2014 from http://pm. lsmuni. lt/foreign/PBL% 20students% 20book. pdf.

Kirkpatrick, D. L. (1998). *Another Look at Evaluating Training Programs*. Alexandria, VA: American Society for Training & Development.

Kirschner, P. A., Sweller, J. & Clark, R. E. (2006). Why minimal guidance during instruction does not work: An analysis of the failure of constructivist, discovery, problem-based, experiential, and inquiry-based teaching. *Educational Psychologist*, 41 (2), 75-86.

Knight, P. T. (2001). Complexity and curriculum: A process approach to curriculum making. *Teaching and Learning in Higher Education*, (6), 369-381.

Kolmos, A. (2007). History of problem-based and project-based learning. In E. Graaff & A. Kolmos (eds.). *Management of Change: Implementation Problem-Based and Project-Based Learning in Engineering* (pp. 1-8). Rotterdam/Taipei: Sense Publishers.

Kolmos, A. (2009). Problem-based and project-based learning: Institutional and global change. In O. Skovsmose, P. Valero & O. R. Christensen (eds.), *University Science and Mathematics*

Education in Transition (pp. 261-280). New York: Springer Science +Business Media LLC.

Kolodner, J. L., Camp, P. J., Crismond, D., Fasse, B., Gray, J., Holbrook, J., *et al.* (2003). Problem-based learning meets case-based reasoning in the middle-school science classroom: Putting Learning by Design™ into practice. *The Journal of the Learning Sciences*, 12 (4), 495-547.

Larsson, J. (2001). Problem-based learning: A possible approach to language education? Retrieved January 15, 2014 from http:// www. nada. kth. se/ ~jla/docs/PBL. pdf.

Laurenceau, J. & Barrett, L. F. (2005). *Journal of Family Psychology*, 19 (2), 314-323.

Leedy, P. (1989). *Practical Research: Planning and Design* (Fourth Edition). New York: Macmillan.

Lesznyak, M. (2007). Conceptualizing translation competence. *Across Languages and Cultures*, 8 (2): 167-194.

Lieux, E. M. (2001). A skeptic's look at PBL. In J. B. Duch, E. S. Groh & E. D. Allen (eds.) (pp. 223-235).

Locke, N. A. (2005). In-house or freelance? A translator's view. *Translation: The Guide from Multilingual Computing and Technology*, 69 Supplement, 19-21.

Luo, X. J. (骆雪娟) (2010). Zhongshan Daxue yingyu zhuanye benkesheng CAT kecheng sheji (中山大学英语专业本科生 CAT 课程设计, "The course design of CAT for undergraduate English majors of Sun Yat-sen University"). *Fanyi Xuebao* (翻译学报, "Journal of Translation Studies"), (13), 252-272.

Luo, X. J. (骆雪娟) (2013). Jiyu jisuanji fuzhu fanyi kecheng jianshe de shichang diaocha (基于计算机辅助翻译课程建设的市场调查, "A market survey in view of computer-aided translation course development"). In J. Y. Xiao (ed.), *Waiyu yu Fanyi Luntan* (外语与翻译论坛, "Language and Translation Forum") (pp. 39-50). Guangzhou: Sun Yat-sen University Press.

Lv, L. S. (吕立松) & Mu, L. (穆雷) (2007). Jisuanji fuzhu fanyi jishu yu fanyi jiaoxue (计算机辅助翻译技术与翻译教学, "CAT technology and translation teaching"). *Waiyujie* (外语界, "Foreign Language World"), (3), 35-43.

Macdonald, R. (2004). Researching the student experience to bring about improvements in problem-based learning. In M. Savin-Baden & K. Wilkie (eds.), *Challenging Research into Problem-based Learning* (pp. 37-49). Berkshire: Society for Research into Higher Education & Open University Press.

Macken, L. (2010). Sub-sentential alignment of translational correspondences. Retrieved April 3, 2013, from https: // www. google. com. hk/url? sa =t&rct =j&q =&esrc =s&source = web&cd =6&ved =0CEAQFjAF&url =http% 3a% 2f% 2fwww% 2felt3% 2feugent% 2febe% 2fmedia% 2fuploads% 2fpublications% 2f2010% 2fMacken2010b% 2epdf&ei =8cAyVMnTIpLv8gW5r4 CICg&usg =AFQjCNF9GrB7JU_zoZigMU2K_bIS_t8Gog&sig2 = Lh3VeCT5zl VHclg81cLxQg.

Macrae, R. P. (2000). *A Problem-Based Approach to Training Private School Administrators about New Faculty Development: A Research and Development Project.* Unpublished Doctor of Education Thesis, Teachers College, Columbia University.

Márta, L. (2008) . *Studies in the Development of Translation Competence.* Unpublished PhD Dissertation. Retrieved December 12, 2012, from http://nydi. btk. pte. hu/sites/nydi. btk. pte. hu/ files/pdf/LesznyakMarta2009_2. pdf.

Masek, A. (2010). Problem-based learning model: A collection from the literature. *Asian Social Science,* 6 (8), 148-156.

Maudsley, G. (1999). Do we all mean the same thing by "problem-based learning"? A review of the concepts and a formulation of the ground rules. *Academic Medicine,* 74 (2), 178-185.

McMillan, M. A. & Dwyer, J. (1989). Changing times, changing paradigm (2): The MacArthur experience. *Nurse Education Today,* (9), 93-99.

Melby, K. (1998). Eight types of translation technology. Paper presented at the American Translators Association (ATA) 39th Annual Conference. November 4-7, 1998, Hilton Head Island, South Carolina, the United States. Retrieved July 18, 2012 from http://www. ttt. org/technology/8types. pdf.

Merriam, S. B. (1998). *Qualitative Research and Case Study Applications in Education.* San Francisco: Jossey-Bass.

Meyers, N. M. & Nulty, D. D. (2009). How to use (five) curriculum design principles to align authentic learning environments, assessment, students' approach to thinking and learning outcomes. *Assessment and Evaluation in Higher Education*, (34), 565-577.

Miao, J. (苗菊) (2007). Fanyi nengli yanjiu: Jian'gou fanyi jiaoxue moshi de jichu (翻译能力研究: 构建翻译教学模式的基础, "Studies on translation competence: Basis for process-oriented translation pedagogy"). *Waiyu yu Waiyu Jiaoxue* (外语与外语教学, "Foreign Languages and Their Teaching"), (4), 47-50.

Miao, J. (苗菊) & Liu, Y. C. (刘艳春) (2010). Fanyi shizheng yanjiu: lilun, fangfa yu fazhan (翻译实证研究: 理论、方法与发展, "Empirical research in translation studies: Theory, methodology and advance"). *Zhongguo Waiyu* (中国外语, "Foreign Languages in China"), (6), 92-97.

Miflin, B. M. (2004). Adult learning, self-directed learning and problem-based learning: Deconstructing the connections. *Teaching in Higher Education*, (9), 43-53.

Miflin, B. M. & Price, D. (2000). Why does the Department have professors if they don't teach? In P. Schwartz, S. Mennin & G. Webb (eds.), *Problem-Based Learning: Case Studies, Experiences and Practice*. London: Kogan Page.

Miles, M. B. & Huberman. M. A. (1994). *Qualitative Data Analysis: An Expanded Sourcebook* (Second Edition). Thousand Oaks, Calif. : Sage.

Miller, H. A., Imrie, W. B. & Cox, K. (1998). *Student Assessment in Higher Education: A Handbook for Assessing Performance*. London: Routledge.

Moore, S. T., Lapan, D. S. & Quartaroli, T. M. (2012). Case study research. In D. S. Lapan, T. M. Quartaroli & F. J. Riemer (eds.) (pp. 243-270).

Moss, P. A. (1992). Shifting conceptions of validity in educational measurement: Implications for performance assessment. *Review of Educational Research*, (62), 229-258.

Mossop, B. (2003). What should be taught at translation schools? In A. Pym, C. Fallada, J. R. Biau & J. Orenstein (eds.), *Innovation and E-learning in Translator Training* (pp. 20-22). Retrieved on

July 10, 2012 from http://isg. urv. es/library/papers/innovation _ book. pdf.

Moust, J. H. C., van Berkel, H. J. M. & Schmidt, H. G. (2005). Signs of erosion: Reflections on three decades of problem-based learning at Maastricht University. *Higher Education*, (50), 665-683.

Mu, L. (穆雷) & Lan, H. J. (蓝红军) (2011). 2010 nian zhongguo fanyi yanjiu zongshu (2010 年中国翻译研究综述, "A review of translation studies in China in 2010"). *Shanghai Fanyi* (上海翻译, "Shanghai Journal of Translators"), (3), 23-28.

Namey, E., Guest, G., Thairu, L. & Johnson, L. (2007). Data Reduction techniques for large qualitative data sets. In G. Guest & K. MacQueen (eds.) *Handbook for Team-based Qualitative Research* (pp. 137-163). Lanham, MD: AltaMira Press.

National Commission for the Protection of Human Subjects of Biomedical and Behavioral Research. (1979) . *The Belmont report: Ethical principles and guidelines for the protection of human subjects of research* (DHEW Publication No. OS 78-0012). Washington DC: Government Printing Office.

Newman, M. J. (2005). Problem based learning: An introduction and overview of the key features of the approach. *Journal of Veterinary Medical Education*, 32 (1), 12-20.

Oberski, I. M., Matthews-Smith, G, Gray, M. & Carter, E. D. (2004). Assessing problem-based learning with practice portfolios: One innovation too many? *Innovations in Education and Teaching International*, 41 (2), 207-221.

Olvera-Lobo, M. D., Robinson, B., Castro-Prieto, R. M., Gervilla, E. Q., Martín, R. M., Raya, E. M. *et al.* (2007). A professional approach to translator training (PATT). *Meta: Translators' Journal*, 52 (3): 517-528.

Onwuegbuzie, J. A., Leech, L. N. & Collins, M. T. K. (2012). Qualitative analysis techniques for the review of the literature. *The Qualitative Report*, (17), 1-28.

Oxford, R. (1997). Constructivism: Shape-shifting, substance, and teacher education applications. *Peabody Journal of Education*, (72), 35-66.

PACTE (1998). Acquiring translation competence: Hypotheses and methodological problems of a research project. In A. Beeby, D. Ensinger, & M. Presas (eds.), *Investigating Translation: Selected Papers from the 4th International Congress on Translation, Barcelona, 1998* (pp. 99-106). Amsterdam/Philadelphia: John Benjamins.

PACTE (2003). Building a translation competence model. In F. Alves (ed.), *Triangulating Translation: Perspectives in Process Oriented Research* (pp. 43-66). Amsterdam/ Philadelphia: John Benjamins.

PACTE (2005). Investigating translation competence: Conceptual and methodological issues. *Meta: Translators' Journal,* (50), 609-619.

PACTE (2009). Results of the validation of the PACTE translation competence model: Acceptability and decision making. *Across Languages and Cultures,* (10), 207-230.

PACTE (2011). Results of the validation of the PACTE translation competence model: Translation project and dynamic translation index. In S. O'Brien (ed.), *IATIS Yearbook 2010*. London: Continuum. Retrieved on March 25, 2012 from http:// grupsderecerca. uab. cat/pacte/sites/grupsderecerca. uab. cat. pacte/ files/2011_PACTE_Continuum. pdf.

Papinczak, T., Young, L. & Groves, M. (2007). Peer assessment in problem-based learning: A qualitative study. *Advances in Health Sciences Education,* 12 (2), 169-186.

Perkins, D. N., & Grotzer, T. A. (2000). Models and moves: Focusing on dimensions of causal complexity to achieve deeper scientific understanding. Paper presented at the American Educational Research Association annual conference, April 2000, New Orleans, LA (ERIC Document Reproduction Service No. ED 441 698).

Prince, M. J. & Felder, R. M. (2006). Inductive teaching and learning methods: Definitions, comparisons, and research bases. *Journal of Engineering Education,* 95 (2), 123-138.

Pym, A. (2003). Redefining translation competence in an electronic age: In defense of a minimalist approach. *Meta: Translators' Journal,* (4), 481-497.

Pym, A. (2009). Translator training. Retrieved July 10, 2012, from http://usuaris. tinet. cat/apym/on-line/training/2009_translator_ training. pdf.

Pym, A. (2012). Translation skill-sets in a machine-translation age. Retrieved August 17, 2012, from http://usuaris. tinet. cat/apym/ on-line/training/2012_competence_pym. pdf.

Qian, C. H. (钱春花) (2011). Jiyu zhagen lilun de yizhe fanyi nengli tixi yanjiu (基于扎根理论的译者翻译能力体系研究, "Study on translation competence: Based on grounded theory"). *Waiyu yu Waiyu Jiaoxue* (外语与外语教学, "Foreign Languages and Their Teaching"), (6), 65-69.

Qian, D. X. (钱多秀) (2009). "Jisuanji fuzhu fanyi" kecheng jiaoxue sikao ("计算机辅助翻译"课程教学思考, "Pedagogical reflections on the design of a course in computer-aided translation"). *Zhongguo Fanyi* (中国翻译, "Chinese Translators Journal"), (4), 51-53.

Qian, D. X. (钱多秀) (2011). *Jisuanji Fuzhu Fanyi* (计算机辅助翻译, "Computer-aided Translation: A Course Book"). Beijing: Foreign Language Teaching and Research Press.

Quah, C. K. (2008). *Translation and Technology*. Shanghai: Shanghai Foreign Language Education Press.

Quintana, C., Reiser, B. J., Davis, E. A., Krajcik, J., Fretz, E., Duncan, R. G., Kyza, E. *et al*. (2004). A scaffolding design framework for software to support science inquiry. *Journal of the Learning Sciences*, (13), 337-386.

Ramsden, P. (1992). *Learning to Teach in Higher Education*. London: Routledge.

Reiв, K. (2000). Type, kind and individuality of text: Decision making in translation. In L. Venuti (ed.) *The Translation Studies Reader* (pp. 160-171). London: Routledge. Reprinted from 1981.

Reynolds, F. (1997). Studying psychology at degree level: Would problem-based learning enhance students' experiences? *Studies in Higher Education*, 22 (3), 263-275.

Robinson, D. & Reed, V. (eds.) (1998). *The A-Z of Social Research Jargon*. Aldershot, UK: Ashgate.

Rodrigo, E. Y. (ed.) (2008). *Topics in Language Resources for*

Translation and Localisation. Amsterdam/Philadelphia: John Benjamins.

Rogers, E. M. (1995). *Diffusion of Innovations*. New York: The Free Press.

Rohani, M. J., Hassan, A., Hassan, S., Hamid, A. & Yusof, K. (2005). Assessing the effectiveness of Problem Based Learning (PBL) using Quality Function Deployment (QFD): Students perspective. *Proceedings of the 2005 Regional Conference on Engineering Education* (pp. 1-6). December 12-13, 2005, Johor, Malaysia.

Ross, B. (1991). Towards a framework for problem-based curricula. In D. Boud & G. Feletti (eds.) (pp. 34-41).

Russell, A. L., Creedy, D. & Davis, J. (1994). The use of contract learning in PBL. In S. E. Chen, S. E. Cowdroy, A. J. Kingsland & M. J. Ostwald (eds.) *Reflections on Problem Based Learning* (pp. 57-72). Sydney: Australian Problem Based Network.

Sager, J. C. (1994). *Language Engineering and Translation: Consequences of Automation*. Amsterdam/Philadelphia: John Benjamins.

Samford University (2003). Problem-based learning at Stamford University. Retrieved on March 23, 2014 from http://www. Samford. edu/pbl/definition. html.

Samson, R. (2010). Computer-aided translation. In M. Tennent (ed.) (pp. 101-126).

Sánchez, M. P. (2006). Electronic tools for translators in the 21st century. *Translation Journal*, (4). Retrieved August 11, 2012 from http://translationjournal. net/journal/38 tools. htm.

Sánchez-Gijón, P., Aguilar-Amat A, Mesa-Lao B. & Solé M. P. (2009). Applying terminology knowledge to translation: Problem-based learning for a degree in translation and interpreting. *Terminology* 15 (1), 105-118.

Savery, J. R. (2006). Overview of problem-based learning: Definitions and distinctions. *Interdisciplinary Journal of Problem-based Learning,* 1 (1), 9-20.

Savery, J. R. & Duffy, T. M. (1995). Problem-based Learning: An instructional model and its constructivist framework. *Educational*

Technology, (5), 31-38.

Savin-Baden, M. (2000). *Problem-based Learning in Higher Education: Untold Stories*. Buckingham: Open University Press/ SRHE.

Savin-Baden, M. (2003). *Facilitating Problem-based Learning: Illuminating Perspectives*. Birkshire: Society for Research into Higher Education and Open University Press.

Savin-Baden, M. (2004). Understanding the impact of assessment on students in problem-based learning. *Innovations in Education and Teaching International*, 41 (2), 223-233.

Savin-Baden, M. & Howell, M. C. (2004). *Foundations of Problem-Based Learning*. Maidenhead, England: Open University Press.

Schäler, R. (1998). The problem with machine translation. In L. Bowker, M. Cronin, D. Kenny & J. Pearson (eds.) *Unity in Diversity? Current Trends in Translation Studies* (pp. 11-156). Manchester: St. Jerome.

Schensul, J. J. (2012). Methodology, methds, and tools in qualitative research. In D. S. Lapan, T. M. Quartaroli & F. J. Riemer (eds.) *Qualitative Research: An Introduction to Methods and Designs* (pp. 69-103). San Francisco: Jossey-Bass.

Schmidt, H. G. (1983). Problem-based learning: Rationale and description. *Medical Education*, (17), 11-16.

Schmidt, H. G. (2000). Assumptions underlying self-directed learning may be false. *Medical Education*, (34), 243-245.

Schmidt, H. G., Boshuizen, H. P. A. & de Vries, M. (1992). Comparing problem-based with conventional education: A review of the University of Limburg medical school experiment. *Ann. Commun.-Oriented Educ.*, (5), 193-198.

Schmidt, H. G., Rotgans, J. I. & Yew, H. J. E. (2011). The process of problem-based learning: What works and why. *Medical Education*, (45): 792-806.

Seltzer, S., Hilbert, S., Maceli, J., Robinson, E. & Schwartz, D. (1996). An active approach to calculus. In L. Wikerson & W. H. Gijselaers (eds.), *Bringing Problem-based Learning to Higher Education: Theory and Practice* (pp. 83-90). San Francisco: Jossey-Bass.

Sharoff, S. (2006). Translation as problem-solving: Uses of comparable corpora. In *Proceedings of Third International Workshop on Language Resources for Translation Work, Research & Training at LREC 2006*, May, 2006, Genoa. Retrieved April 2, 2014 from http://corpus. leeds. ac. uk/serge/ publications/lrec2006-lr4trans. pdf.

Shi, Z. L. (史宗玲) (2004). *Diannao Fuzhu Fanyi* (电脑辅助翻译, "MT & TM"). Taiwan: Bookman Books Co., Ltd.

Shipman, L. H. & Duch, J. B. (2001). Problem-based learning in large and very large classes. In J. B. Duch, E. S. Groh & E. D. Allen (eds.) (pp. 149-163).

Shor, I. & Freire, P. (1987). *A Pedagogy for Liberation*. London: Bergin and Garvey.

Silver, C., Lewins, A. & Patashnick, J. (2014). NVivo. In Silver, C. & Lewins, A. (eds.) *Using Software in Qualitative Research: A Step-by-Step Guide* (2nd Edition). London: Sage. Retrieved on November 12, 2013 from https: //study. sagepub. com/using-software-in-qualitative-research/student-resources/step-by-step-software-guides/nvivo-10.

Simons, K. D. & Ertmer, P. A. (2005). Scaffolding disciplined inquiry in problem-based environments. *International Journal of Learning*, 12 (6), 297-305.

Slavin, R. (1990). *Cooperative Learning: Theory, Research and Practice*. Boston, MA: Allyn & Bacon.

Somers, H. (ed.) (2003). *Computers and Translation: A Translator's Guide*. Amsterdam/Philadelphia: John Benjamins.

Springer, L., Stanne, M. E. & Donovan, S. S. (1999). Effects of small-group learning on undergraduates in science, mathematics, engineering, and technology: A meta-analysis. *Review of Educational Research*, 69 (1), 21-51.

Stake, R. (1995). *The Art of Case Study Research*. Thousand Oaks, London, New Delhi: Sage.

Stake, R. (1998). Case Studies. In N. K. Denzin & Y. Lincoln (eds.), *Strategies of Qualitative Inquiry*. Thousand Oaks, London, New Delhi: Sage.

Stake, R. (2005) . Qualitative Case Studies. In N. K. Denzin & Y. S.

Lincoln (eds.). *The Sage Handbook of Qualitative Research* (3rd Ed). Thousand Oaks, London: Sage.

Steele, D. J., Medder, J. & Turner, D. P. (2000). A comparison of learning outcomes and attitudes in student-versus faculty-led problem-based learning: An experimental study. *Medical Education*, (34), 23-29.

Stewart, J., Orbán, W. & Kornelius, J. (2010). Cooperative translation in the paradigm of problem-based learning. *T21N — Translation in Transition*, (1), 1-28.

Stupiello, N. A. (2008). Ethical implications of translation technologies. *Translation Journal*, (12). Retrieved July 12, 2012 from http://www. bokorlang. com/journal/43ethics. htm.

Syllabus of CAT 2, (2012). CA: Monterey Institute of International Studies.

Tai, G. X. & Yuan M. C. (2007). Authentic assessment strategies in problem based learning. In *Proceedings ICT: Providing Choices for Learners and Learning* (pp. 983-993). December 2-5, 2007, Nanyang Technological University, Singapore.

Tan, Oon-Seng (2007). Problem-based learning pedagogies: Psychological processes and enhancement of intelligences. *Educational Research for Policy and Practice*, (6), 101-114.

Taylor, C. (1994). Assessment for measurement or standards: The peril and promise of large-scale assessment reform. *American Educational Research Journal*, (31), 231-262.

Taylor, D. & Miflin, B. (2008). Problem-based learning: Where are we now? *Medical Teacher*, (30), 742-763.

Tiantong, M. & Teemuangsai, S. (2013). The four scaffolding modules for collaborative problem-based learning through the computer network on moodle LMS for the computer programming course. *International Education Studies*, 6 (5), 47-55.

Tong, L. C. (1994). Translation: machine-aided. In R. E. Asher & J. M. Y. Simpson (eds.) *The Encyclopedia of Language and Linguistics*, Vol. 9 (pp. 4730-4737). Oxford: Pergamon Press.

Torp, L. & Sage, S. (2002). *Problems as Possibilities: Problem-Based Learning for K-16 Education* (Second Edition). Alexandria, VA: Association for Supervision and Curriculum Development.

Trafton, P. R. & Midgett, C. (2001). Learning through problems: A powerful approach to teaching mathematics. *Teaching Children Mathematics*, 7 (9), 532-536.

Valle, R., Petra, I., Martínez-González, A., Rojas-Ramirez, A. J., Morales-Lopez, S. & Pinña-Garza, B. (1999). Assessment of student performance in problem-based learning tutorial sessions. *Medical Education,* (33), 818-822.

van Lier, L. (2004). *The Ecology and Semiotics of Language Learning.* Dordrecht: Kluwer Academic.

von Glasersfeld, E. (1995). A constructivist approach to teaching. In L. Steffe & J. Gale (eds.) *Constructivism in Education* (pp. 3-16). Mahwah, NJ: Erlbaum.

von Glasersfeld, E. (1998). Cognition, construction of knowledge and teaching. In M. R. Matthews (ed.) *Constructivism in Science Education* (pp. 11-30). London: Kluwer.

Vygotsky, L. (1978). *Mind in Society: The Development of Higher Psychological Processes.* Cambrige, MA: Harvard University Press.

Walqui, A. (2006). Scaffolding instruction for English language learners: A conceptual framework. *The International Journal of Bilingual Education and Bilingualism,* 9 (2), 159-180.

Wang, C. Y. (王传英) (2012). 2011 nian qiye yuyan fuwu rencai xuqiu fenxi ji qishi (2011 年企业语言服务人才需求分析及启示, "Survey on enterprises' needs for language services human resources: Analysis and implications"). *Zhongguo Fanyi* (中国翻译, "Chinese Translators Journal") (1), 67-70.

Wang, H. S. (王华树) (2012). Xinxihua shidai beijing xia de fanyi jishu jiaoxue shijian (信息化时代背景下的翻译技术教学实践 "Translation technology teaching practice in the information age"). *Zhongguo Fanyi* (中国翻译, "Chinese Translators Journal") (3), 57-62.

Wang, H. S. (王华树) (2013). Yuyan fuwu hangye jishu shiyu xia de MTI jishu kecheng tixi goujian (语言服务行业技术视域下的 MTI 技术课程体系构建, "Construction of MTI technology curriculum in the context of technological development of the language service industry"). *Zhongguo Fanyi* (中国翻译, "Chinese Translators Journal") (1), 23-28.

Wang, L. D. (王立弟) (2012). A more inclusive model of translator and interpreter training. In H. Lee-Jahnke, M. Forstner & Lidi Wang (eds.), *A Global Vision: Development of Translation and Interpreting Training: Proceedings of CIUTI-Forum Beijing 2011* (pp. 55-65). Beijing: Foreign Language Teaching and Research Press.

Wang, S. H. (王树槐) & Wang, R. W. (王若维) (2008). Fanyi nengli de goucheng yinsu he fazhan cengci yanjiu (翻译能力的构成因素和发展层次研究, "On the components and developments of translation competence"). *Waiyu Yanjiu* (外语研究, "Foreign Languages Research") (5), 80-88.

Wang, T. (王涛) & Lu, P. (鹿鹏) (2008). Fanyi jishu de linian yu fenlei (翻译技术的理念与分类, "The concepts and classifications of translation technology"). *Zhongguo Keji Fanyi* (中国科技翻译, "Chinese Science & Technology Translators Journal") (1), 20-23.

Wang, X. L. (王湘玲), Mo, J. (莫姣) & Liang, P. (梁萍) (2010). Dianzi Fanyi gongju yu EFL xuesheng Fanyi gongju nengli zhi peiyang (电子翻译工具与 EFL 学生翻译工具能力之培养, "Electronic translation tools and development of EFL students' instrumental competence"). *Waiyu yu Waiyu Jiaoxue* (外语与外语教学, "Foreign Languages and Their Teaching"), (6), 75-78.

Wang, X. L. (王湘玲), Tang, W. (汤伟) & Wang, Z. M. (王志敏) (2008). Xifang Fanyi nengli yanjiu: Huimou yu qianzhan (西方翻译能力研究：回眸与前瞻, "Research on translation competence: Past and future"). *Hunan Daxue Xuebao (shehui kexue ban)* [湖南大学学报 (社会科学版), "Journal of Hunan University (Social Sciences)"] (2), 103-106.

Weaver, W. (1949). Translation. Repr. in W. N. Locke & A. D. Booth (eds.) *Machine Translation of Languages: Fourteen Essays* (pp. 15-23). Cambridge, Mass.: Technology Press of the Massachusetts Institute of Technology, 1955.

Wee, L. N. K. (2004). A problem-based learning approach in entrepreneurship education: Promoting authentic entrepreneurial learning. *International Journal of Technology Management*, 28 (7/8), 685-701.

Wen, J. (文军) (2005). *Fanyi kecheng moshi yanjiu: Yi fazhan Fanyi nengli wei zhongxin de fangfa* (翻译课程模式研究：以发展翻译能力为中心的方法, "Study of translation curriculum models: Translation-competence oriented curriculum model"). Beijing: Chinese Literature and History Press.

Wen, J. (文军) & Li, H. X. (李红霞) (2010). Yi Fanyi nengli wei zhongxin de Fanyi zhuanye benke kecheng shezhi yanjiu (以翻译能力为中心的翻译专业本科课程设置研究, "A study on the translation competence-centered curriculum design for translation major"). *Waiyujie* (外语界, "Foreign Language World"), (2), 2-7.

Widdowson, H. G. (1984). *Explorations in Applied Linguistics 2*. Retrieved August 24, 2012 from http://bearsite. info/Articles/Language/Linguistics% 20Texts/Explorations% 20in% 20Applied% 20Linguistics% 202/.

Wilkerson, L. and Hundert, E. M. (1991). Becoming a problem-based tutor: Increasing self-awareness through faculty development. In D. Boud & G. Feletti (eds.) (pp. 160-172).

Wilss, W. (1996). *Knowledge and Skills in Translator Behaivor*. Amsterdam/Philadelphia: John Benjamins.

Wood, (1988). *How Children Think and Learn*. Oxford, Basil Blackwell.

Wood, F. D. (2003). ABC of learning and teaching in medicine: Problem-based learning. *British Medical Journal*, 326, 328-330.

Woods, D. R. (1994). *Problem-based Learning: How to Gain the Most from PBL*. Waterdown, Canada: Woods.

Woods, D. R. (1995). *Problem-based Learning: Helping Your Students Gain the Most from PBL*. Hamilton, Ontario: Donald R Woods.

Woods, D. R. (1996). Problem-based learning for large classes in chemical engineering. In L. Wilkerson and H. Gijselaers (eds.) *Bringing Problem-based Learning to Higher Education: Theory and Practice* (pp. 91-99). San Francisco, CA: Jossey-Bass.

Woods, D. R. (2000). Helping your students gain the most from PBL. In O. S. Tan *et al.* (eds.) *Problem-based Learning: Educational Innovation across Disciplines*. Singapore: Temasek Centre for Problem-based Learning.

Xu, B. (徐彬) (2006). CAT yu Fanyi yanjiu he jiaoxue (CAT 与翻译研究和教学, "Application of CAT in teaching and researching"). *Shanghai Fanyi* (上海翻译, "Shanghai Journal of Translators") (4), 59-63.

Xu, B. (徐彬) (2010a). Jisuanji fuzhu Fanyi jiaoxue: sheji yu shishi (计算机辅助翻译教学: 设计与实施, "Computer-aided translation teaching: Design and implementation"). *Shanghai Fanyi* (上海翻译, "Shanghai Journal of Translators") (4), 45-49.

Xu, B. (徐彬) (2010b). *CAT: A New Horizon for Translating Research and Practice*. Jinan: Shandong Education Press.

Xu, B. (徐彬), Guo, H. M. (郭红梅) & Guo, X. L. (国晓立) (2007). 21 shiji de jisuanji fuzhu Fanyi gongju (21 世纪的计算机辅助翻译工具, "Applications of computer-aided translation: An overview"). *Shandong Waiyu Jiaoxue* (山东外语教学, "Shandong Foreign Language Teaching Journal") (4), 79-86.

Yamane, D. (2006). Concept preparation assignments: A strategy for creating discussion-based courses. *Teaching Sociology*, (34): 236-248.

Yuan, Y. N. (袁亦宁) (2005). Fanyi jishu yu woguo jishu Fanyi rencai de peiyang (翻译技术与我国技术翻译人才的培养, "Translation technology and training of technical translators in China"). *Zhongguo Keji Fanyi* (中国科技翻译, "Chinese Science & Technology Translators Journal") (1), 51-54.

Zhang, X. J. (张霄军) (2010). Yingguo gaoxiao de "fanyi jishu" jiaoxue jiqi qishi (英国高校的 "翻译技术" 教学及其启示, "Translation technology teaching in the UK universities and their pedagogical implications"). *Waiyu Yanjiu* (外语研究, "Foreign Languages Research") (6), 76-79.

Zheng, Y. (郑晔) & Mu, L. (穆雷) (2007). Jin 50 nian Zhongguo Fanyi jiaoxue yanjiu de fazhan yu xianzhuang (近 50 年中国翻译教学研究的发展与现状, "A review of research of translation teaching in the mainland of China in the past 50 years"). *Guangdong Waiyu Waimao Daxue Xuebao* (广东外语外贸大学学报, "Journal of Guangdong University of Foreign Studies") (5), 60-66.

Zhong, W. H. (仲伟合) (2007). Fanyi shuoshi zhuanye xuewei

(MTI) ji qi dui Zhongguo waiyu jiaoxue de tiaozhan (翻译硕士专业学位（MTI）及其对中国外语教学的挑战, "Opportunity and challenges: An interpretation of the newly established professional degree programme — master of translation and interpreting"). *Zhongguo Waiyu* (中国外语, "Foreign Languages in China") (4), 4-12.

Zhong, W. H. (仲伟合) (2011). Gaodeng xuexiao Fanyi zhuanye benke jiaoxue yaoqiu (高等学校翻译专业本科教学要求). *Zhongguo Fanyi* (中国翻译, "Chinese Translators Journal") (3), 20-24.

Zhou, X. H. (周兴华) (2013). Jisuanji fuzhu Fanyi jiaoxue: Fangfa yu ziyuan (计算机辅助翻译教学：方法与资源, "Computer-aided translation teaching: Methods and resources"). *Zhongguo Fanyi* (中国翻译, "Chinese Translators Journal") (4), 91-95.

Net 1. UCL, http://www. ucl. ac. uk/centras/scientific-technical-medical-translation-msc/coremodules. [Retrieved on May 19, 2013]

Net 2. SOAS, http://www. soas. ac. uk/courseunits/15PLIC018. html. [Retrieved on May 19, 2013]

Net 3. Dublin University, http://www. dcu. ie/registry/module_contents. php? function =2&subcode = LC501. [Retrieved on May 19, 2013]

Net 4. School of Cultures, Languages and Area Studies in Nottingham University (First phase), http://modulecatalogue. nottingham. ac. uk/Nottingham/asp/ModuleDetails. asp? crs_id =024556&year_id = 000113. [Retrieved on May 19, 2013]

Net 5. School of Cultures, Languages and Area Studies in Nottingham University (Second phase), http://modulecatalogue. nottingham. ac. uk/Nottingham/asp/ModuleDetails. asp? crs_id =024557&year_id =000113. [Retrieved on May 19, 2013]

Net 6. Department of Languages, Translation and Communication in Swansea University, http://www. swansea. ac. uk/media/Enrolment% 20Guide% 20and% 20Handbook% 20for% 20MA% 20Translation% 20Students. pdf. [Retrieved on May 19, 2013]

Net 7. Course Design through Constructive Alignment. Retrieved on July 18, 2014, from, http://www. coles. uoguelph. ca/pdf/Course% 20Design% 20Handout% 20-% 202. pdf. [Retrieved on July 18, 2014]

Appendices

Appendix 1 Participants and Grouping Information

Team One (Best gender structure)				
Student No.	Name	GPA	Gender	Note
11307 * * *	郭 × ×	4.0	Female	/
11307 * * *	洪 ×	2.9	Female	/
11307 * * *	张 × ×	3.8	Female	Strong motivation; great interest in pursuing translation as a profession
11307 * * *	张 × ×	3.4	Male	/
11307 * * *	丛 × ×	3.3	Male	/
Team Two (Most varied motivations)				
Student No.	Name	GPA	Gender	Interest in translation as a profession
11307 * * *	罗 × ×	3.9	Female	Strong
11307 * * *	尹 × ×	3.9	Female	Strong
11307 * * *	郑 ×	3.8	Male	Low
11307 * * *	张 × ×	3.4	Female	Very Strong
11307 * * *	林 × ×	3.1	Female	Low
Team Three (Academic performances evenly distributed)				
Student No.	Name	GPA	Gender	Note
11307 * * *	刘 × ×	3.2	Female	/
11307 * * *	王 × ×	3.0	Male	Easy-going and humorous
11307 * * *	籍 × ×	3.6	Female	/

Continued

Team Three (Academic performances evenly distributed)				
Student No.	Name	GPA	Gender	Note
11307 * * *	陆 × ×	3.8	Female	/
11307 * * *	朱 × ×	4.0	Female	/
Team Four (Most similar academic performances)				
Student No.	Name	GPA	Gender	Different Prior Knowledge/Skills
11307 * * *	白 × ×	3.7	Female	With prior experience in translation and strong desire to be a translator
11307 * * *	张 × ×	3.6	Female	/
11307 * * *	李 × ×	3.6	Male	The only one with prior knowledge in software and programming and showing intense interest in natural science
11307 * * *	汤 × ×	3.8	Female	/
11307 * * *	罗 × ×	3.8	Female	/
Team Five (Single gender)				
Student No.	Name	GPA	Gender	Note
11307 * * *	肖 × ×	3.9	Female	/
11307 * * *	刘 × ×	3.6	Female	/
11307 * * *	王 × ×	3.5	Female	/
11307 * * *	郑 × ×	3.4	Female	/
11307 * * *	罗 × ×	/	Female	Dropped half way

Appendix 2　The Questionnaire of Open-ended Questions

答题提示：下面是一些开放式的问题，请根据你在此次课程中的真实感受进行回答。回答无所谓优劣，欢迎畅所欲言，特别希望听到不同的声音和建设性的意见。建议回答尽可能详尽，对于看法和观点要举例加以说明或论证。

Directions: Below you will be expected to answer some questions based on your experience in this CAT course. There are no right-or-wrong answers to these questions. You're welcome and encouraged to share with us your own feelings and thoughts. Different voices and constructive suggestions are particularly wanted. Do remember to provide brief explanation or justification for your opinions and viewpoints.

Stage 1　（Delivered right after the course is over）

1. 无论是开心还是痛苦，请简单谈谈你对此次 CAT 课程采用的 PBL 教学法的感受。How do you think of the new PBL approach to this CAT course? Please give brief reasons for what experience it has bought you, be it happy or painful.

2. 你在这次课程中学到了什么？如果可以，请简要解释一下所学到的内容。What have you learnt in this course? List them and, if possible, try to explain briefly.

3. 你是否有兴趣在课后继续关注、自学 CAT？如果有兴趣，是否有具体的计划去学习些什么呢？Are you interested to further study CAT on your own after the course is over? If yes, have you had any specific idea about what to learn in the future?

4. 你如何评价自己对课程涵盖的各种工具的掌握程度？如果可以，请按"很差""合格""中等""良好""优秀"给自己的能力评级，并请简单说明理由。How do think of your command of the tools covered in this course? Grade your ability on the scale of "poor, pass, satisfactory, good, excellent" and give brief reasons.

5. 你觉得经过此次课程后，你在翻译学习和实践中有没有什么改变？如果有，请列举并加以说明。Do you think this course with

a PBL approach has caused you any changes in translation learning and practice? If yes, what are they? Please explain briefly.

6. 请简单评价一下此次 PBL 教学法中的问题设计、小组学习的方式和教师角色转变的效果。Please comment on the following three components in the PBL approach: The problem design, learning in groups and the role of teachers changed to be facilitators.

7. 你认为此次课程中最困难和最有益的部分各是什么? What did you find is the most difficult and the most rewarding part for you in this course?

8. 你对该课程的还有什么建议吗? What suggestions do you have for the course?

Stage 2 (*Delivered one year after the course is over*)

9. 一年过去了, 你们是否还记得计算机辅助翻译课程的内容呢? 你们从这个课程中了解或习得的任何知识、技能和态度是否有助于过去一年里你们与翻译相关的学习和实践呢? 如果有, 具体是哪些知识、技能和态度, 它们是如何帮助你们的呢? It's been a year since you took the course of Computer-aided Translation last summer. Do you still remember the content of the course? Do you find anything you learnt from the course, be it the knowledge, skills or attitude, useful in your later learning and practice concerning translation? If yes, what is it and how has it helped you?

Appendix 3　Student Individual Journal Template

Student Name: _____　Date: _____

No.	Task Description	Time Span	Tools/Purposes	Resources/ Purposes	Results & Reflections
		__:__ ～ __:__	1. _____ for _____ 2. _____ for _____	1. _____ for _____ 2. _____ for _____	Daily Self-assessment: ( ) points
		__:__ ～ __:__	1. _____ for _____ 2. _____ for _____	1. _____ for _____ 2. _____ for _____	Merits: Demerits:
		__:__ ～ __:__	1. _____ for _____ 2. _____ for _____	1. _____ for _____ 2. _____ for _____	How can I do better?

Appendix 4　Student Group Study Record Template

Group ＿＿　DATE: ＿＿ / ＿＿ / ＿＿　(Y/M/D)　Time: ＿＿ - ＿＿　Place: ＿＿＿

Activity Goals/Type	Goal (s): Form:	Description:			
Steps: Objectives	Tools/Resources Used	Contributor	Contribution	Results/Findings	Scribe
Process　1.					
2.					
3.					
4.					
5.					

273

Continued				Reflections
Self-assessment	Member 1 : Member 2 : Member 3 : Member 4 : Member 5 :	Peer evaluation	Member 1 : Member 2 : Member 3 : Member 4 : Member 5 :	

Appendix 5　Classroom Observation Sheet

Date:　　　　Time:　　　　Place:　　　　Observer:

[No.] Activities	Description	Time Spent			Participant roles		Observer's Notes
		Begin at	End at	Total (Min)	Teacher	Student	
					____ Facilitator ____ Lecturer ____ Tutor ____ Listener *Others*	____ Group work ____ Individual work ____ Class discussion ____ Lecture listener *Others*	
					____ Facilitator ____ Lecturer ____ Tutor ____ Listener *Others*	____ Group work ____ Individual work ____ Class discussion ____ Lecture listener *Others*	

Appendix 6 Structure of the Dropbox Folder

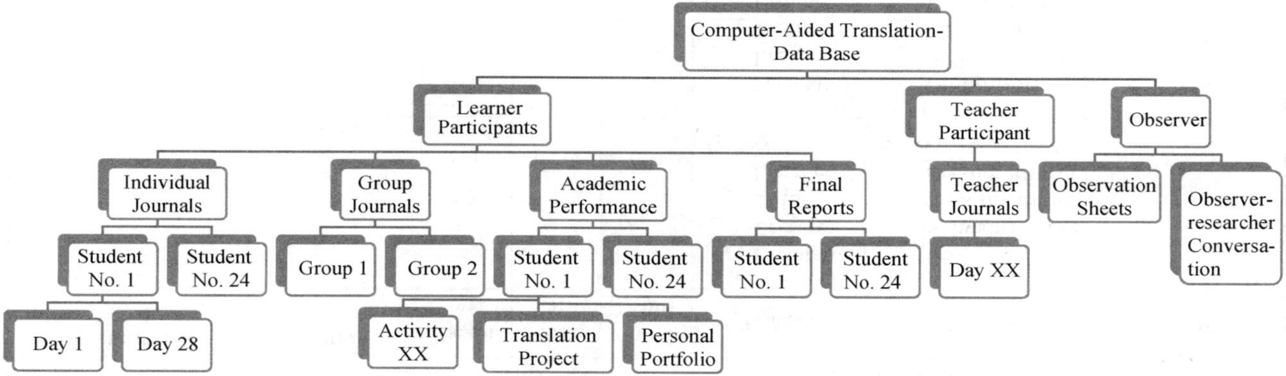

Appendix 7 How to Use the Dropbox Folder

A. Refer to https: //www. dropbox. com/, the official website of Dropbox, and learn how to use it on your own.

B. Apply for an account and install the client terminal on your PC or laptop.

C. Find the folder *Computer-aided Translation* and familiarize yourself with the structure under it.

D. Locate your personal folder that has been created for you and put your files into the specified one as required during the course.

E. Read the following rules and disciplines and keep adhered to it strictly.
   a. Update your *Dropbox* folder as required in time;
   b. Make a copy of all your personal files and those files you are interested in the shared *Dropbox* folder elsewhere on your own computer for the sake of information safety as the fold is always in a dynamic state, being accessible to everybody;
   c. Any changes to the files in the shared *Dropbox* folder other than those under your own name are forbidden;
   d. Do not change the files you have submitted previously unless necessary. Do remember to inform the tutor of the changes via email.

Appendix 8　A Survey on Student Needs

A brief survey was carried out among the participating students to elicit the students' prior knowledge about CAT and what they wanted to learn from the course. Their responses are listed below in the table. The three questions in the survey are：

1. What do you know about translation tools and their functions?
2. What do you expect to learn from this course?
3. Ask whatever questions you have about the course.

Student No.	Questions from the students about the course
1	1. 这门课程的目标是什么？怎么评价翻译能力？ 2. 翻译工具好坏的标准是什么？
2	老师您认为翻译工具的发展方向是应该更加标准化还是更加人性化？
3	1. 利用搜索功能时，什么是关键字组合和语法组合？ 2. 学习或使用一些专业的翻译工具时，需要掌握某些专业的计算机技术吗？
4	1. 较常用、较好用的翻译工具有哪些？ 2. 翻译工具的使用能必然提高翻译效率吗？存在弊端吗？
5	做翻译除了要迅速找到自己需要的信息（在网上），也应该学会如何积累和整理自己已做过的翻译中有用的信息。比方说术语、固定语法等。这种整理已积累信息的能力怎么培养？它算是翻译工具能力吗？
6	在多大程度上或在哪些方面翻译工具可以代替人工翻译？过分依赖翻译工具会不会使译者自身能力退化？翻译工具可否用于翻译外的其他领域的学习？
7	1. 我想知道什么是 back-translation？ 2. 翻译工具有哪些类型，像是图书馆、词典、软件？ 3. 如何在使用翻译工具的同时保持所翻译的 fidelity？ 4. 在搜索翻译工具的时候发现一个专有名词叫作互操作性，是有什么作用的？

Continued

Student No.	Questions from the students about the course
8	1. 以后的翻译流程是否会出现"读文本→机器翻译→调整→定稿"的模式？ 2. 有没有专业性强的翻译工具？比如我想翻译一篇天文学方面的深度文献，除了有道等普通工具外，有没有更权威、更有针对性的工具？ 3. 工具发展是否意味着对翻译从业人员专业素质要求降低？我们还可以从哪些方面提高专业素养？
9	利用翻译工具辅助翻译有什么专业性的步骤和技巧？怎样选择翻译工具？有什么依据区分各个翻译工具？怎样确保网络资源的正确性？
10	翻译工具的使用会不会让人变懒以至于降低他的英语能力？
11	1. 翻译工具的学习等同于 CAT 的学习吗？ 2. 谷歌翻译这类的在线翻译算得上翻译工具吗？称得上 CAT 的工具吗？ 3. 雅信软件、Trados, Multiterm、SDLX 与 Deja Vu X 的破解版与未破解版有什么差别？
12	1. 我们有必要再去上小学期的其他课程吗？ 2. 该课程是否有利于提高我们信息检索以及处理的能力？
13	对翻译工具了解不深，确实也道不出个所以然来。个人觉得，翻译工具基本上都是按套路原理来的，所以很多东西都是千篇一律。"要提高机译的质量，首先要解决的是语言本身的问题而不是程序设计问题，单靠程序来做机译系统，肯定无法提高机译质量。"对于这句话，不能同意更多。
14	1. 如果翻译过程中遇到一些非英语的外来词，因为自身对此非英语语言的认识不足，可能无法准确把握意思。 2. 翻译过程中难免会遇到一些不懂的术语，缺少相关知识也会导致认识偏差。 3. 汉语和英语有些词汇不能一一对应，或者是新兴流行的词汇，例如"屌丝""狗血"，如何准确翻译成英文？ 4. 地名的译法。
15	1. 这门课侧重于培养使用电子工具的能力，对计算机使用能力会不会有较高要求？ 2. 希望老师能穿插一些关于怎样培养逻辑思维的讲解，也分享一下作为译者的经历。

Continued

Student No.	Questions from the students about the course
16	随着互联网技术的发展，使用翻译工具是否更侧重于利用网络资源来实现翻译过程的快捷准确？随着翻译工具智能化发展，过度依赖翻译工具是否会对翻译从业人员语言能力要求降低呢？
17	1. 在完成一个翻译任务时，我们应该在一个什么样恰当的时候使用翻译工具？ 2. 通常情况下我们所要完成的翻译任务都涉及到一定的专业领域，我们是否需要有一个专业背景和专业词汇系统储存？
18	1. 什么类型的材料适合大量地使用翻译工具？ 2. 在用翻译工具辅助翻译工作时，如果将较长的段落自己先分成几小段，依次用工具翻译，是否会提高翻译的准确性？ 3. 我之前用翻译工具（如 GOOGLE TRANSLATE）翻译时，出来的翻译结果常常很糟糕，我通常会对这些病句进行改错、美化。有的句子较好改，但有些是越改越不对劲。请问这时应该放弃工具翻出的句子自己重翻吗？如果经常遇到这种情况岂不是很浪费时间？
19	这门课主要内容是什么呢？翻译工具对翻译真的有很大帮助吗？
20	1. 在翻译过程中，尤其是文学作品中总会遇到一些作者较为独特的表达与写法（比如说一些令人初次难以理解的句式、比喻、感情等），即使能看懂也很难恰当翻译其情感色彩。又或者自己本身的理解与作者的意思可能南辕北辙，且没有过多的以往的类似资料或翻译材料作比较。此时应该如何评判校对自己的翻译文本？ 2. 在进行专业性较高的翻译时，有必要先用自动翻译工具粗略翻译理解一下大意吗？还是说这种先入为主的行为完全要不得，必须亲自翻。并且在遇到一些专业知识且并非外行人可以简单掌握的时候，仅仅根据句法翻译很可能化简为繁，如何避免这种弄巧成拙？

Continued

Student No.	Questions from the students about the course
21	网站类 1. 当手中有中文（英文）的资料（如论文等有权威的材料），为了能找到其中某 些词汇的可靠翻译以更好地服务相关方面的翻译，应该如何快速在网上找到其平行文本？ 2. 各大搜索引擎各自支持哪些搜索方式以及各有何便利之处？（e.g. Google 支持通配符 * 代表不确定的词，可用于查找相关搭配） 3. 有哪些比较好的、有助于翻译的网站（除百度、谷歌、必应等知名的网站外)？ 电子词典类 4. 平时多用有道、必应电子词典，但这些词典没有专攻的方面。有哪些比较有特点的词典？（e.g. urban dictionary 专门收集俚语等等)
22	这门课程具体学习内容是什么？课程难度大吗？要学好这门课程需要做哪些准备？
23	如何利用网上的语料库资源？翻译语料库与平行语料库、单语语料库有什么异同？
24	什么样的翻译工具才是出色的翻译工具？电子翻译工具在口译领域有帮助吗？如果有，是起到怎样的作用呢？在实践中难运用吗？电子翻译工具对文学作品翻译有何益处？对信达雅会否有较大积极或消极影响？在翻译文学作品时，尽量少使用电子翻译工具会否更好？本地化与翻译工具之间关系如何？如何有效快速从网上检索到有用信息？

Appendix 9　The Problems Design Matrix

No.	Topic	ILOs addressed	Knowledge type	Cognitive level	Format	Class contact (academic hour)
1	CAT and MT technology: What, how and why	ILOs 1 & 2	Declarative	Remembering & Understanding	Textual: Story telling	6
2	Broad-sense CAT: BYU corpora and Google Search	ILOs 4, 5, 7, 8, 10, 11 & 12	Declarative & Functioning	Remembering & Understanding & Applying & Analysing & Evaluating	Translation scenarios	6
3	Narrow-sense CAT: SDL Trados 2011, Oriental Yaxin or Google Translator Toolkit	ILOs 1-12	Declarative & Functioning	Remembering & Understanding & Applying & Analysing & Evaluating & Creating	Translation project	12
4	Translation project management and CAT	ILOs 1-12	Declarative & Functioning	Remembering & Understanding & Applying & Analysing & Evaluating & Creating	Translation project	10

Appendix 10　The Second Problem（Full Version）

Try to solve in groups at least 5 of the following problems (*The more, the better!*), using whatever tools and resources you can find to help you. Online tools and resources are preferable. Do support your opinions with evidence. Remember you'll have to note down what tools and resources are used for each problem. For electronic ones, it would be better to include screenshots.

Tutor's recommendation: *the must-uses*

- Internet search engines: Google and its search Tips & Tricks http://www. google. com/intl/en/insidesearch/tipstricks/
- BYU corpus http://corpus. byu. edu/
- Online dictionaries: e.g. 对比词典 http://www. diffen. com/ etc.
- Online database: e.g. MBA 智库百科 http://wiki. mbalib. com/ wiki/MBA etc.
- Professional resources: e.g. @一本词典 计算机辅助翻译微刊：http://kan. weibo. com/kan/3444217594967632
- Quality bilingual websites: e.g. http://www. who. int/en/

Translation Scenarios：

1. 在一次翻译金融类文章时遇到了 oligopoly 和 monopoly 两个词，不知道两个词有什么差别，应该怎么翻。**如果是你，你会怎么做？**

2. 一个同学有一次翻译有关环境污染的新闻，其中出现"棕地治理"一词，根据上下文，了解到"棕地"是指化工厂搬迁后遗留下来的有污染物的土地，可是查阅了所有词典也没有找到这个词的翻译。**他非常为难，你将如何解决这个词的翻译问题呢？**（这个问题可能有多个解决途径，你可以多尝试一些途径，并比较它们之间的效果差异）

3. 译者在翻译一个小说时，曾遇到"green zebra"一词，从上下文可以猜出是种食物，但是网络上几乎没有相关的中文信息。于是译者询问了在美国的几个同学，大概了解到了它是一种带有绿白色斑马纹的绿番茄，可是最后还是没有找到对应的中文官方名称，就直接译作了"绿番茄"。

请评价这个译者的翻译过程和译文选择。如果是你，你会怎么做？

4. 一个译者在听政府报告时，发现翻译多把国债翻译成 treasury bond，很多场合下的国债的翻译都被译作 treasury bond。有一次在字典中查找 treasury，发现多指政府发行的国库券，而国库券是指不超过一年的短期证券，所以他觉得将国债翻译成 treasury bond 是不准确的，应该使用 national debt。

你如何评价上述的思考过程？如果是你，你会怎么做？

5. 有一次看见在讨论网上书店和实体书店的优劣对比时，一个译者将"实体书店"翻译成"entity bookstores"，entity 确实有"实体"的意思。**那么如何判断这个译法对不对呢？为什么？**

6. "一次翻译中有 Marc Augé 的建筑理念 non-places；我为了找 non-places 的惯常译法，去搜索 Marc Augé 的资料。在百度输入'Marc Augé'，得到马克·奥日，然后输入'马克·奥日 non-places'我找到一些链接是采访马克·奥日的，其中有提及其理念 non-places，译为'他处'；再多点击几个链接，还有译法为'非地'等，我根据文章和个人感觉，选择了'他处'的译法。"

请评价这位译者的做法。

7. 一位同学在参加志愿服务时，有位哥哥让他帮忙翻译"志愿者联合会"这个词，对于"联合会"的翻译，他的脑海里蹦出 federation/union/association 等词，通过查阅词典，他觉得 federation 是政治上的联合，association 好像更常用。

如果是你，你会怎样去确定最后的译法？

8. 有个同学在翻译一本小说，其中有这样一句话：

She felt perversely glad to be in mourning: no one would expect her to behave completely normally at a time like this. Her confusion, distraction, and startled responses to unsaid words didn't seem too terribly out of place, even if they felt cringe — inducingly noticeable to her. (Note：这部小说的女主人公有读心术，所以能听见别人的思想)

学生最初的翻译是：她不合常理地庆幸起自己身处在葬礼上，因为没人会指望她在这种场合，还保持正常得体的表现。她的慌张困惑、心烦意乱和对听见别人的思想惊愕的反应，即使十分明显，也不会被视为不合时宜。

你觉得这个翻译准确吗？理解起来有困难吗？如果有，你会怎样去解决？

9. For the man who is extremely and dangerously hungry, no other interests exist but food. He dreams food, he remembers food, he thinks about food, he emotes only about food, he perceives only food and he wants only food. The more subtle determinants that ordinarily fuse with the physiological drives in organizing even feeding, drinking or sexual behavior, may now be so completely overwhelmed as to allow us to speak at this time (but only at this time) of pure hunger drive and behavior, with the one unqualified aim of relief.

这句话理解起来有困难吗？你会怎样去解决这个困难？请给出你的译文。

10. 在一次英文写作中，你想表达这句话：我希望通过加大阅读量来让自己的注意力更加集中。你觉得这样的翻译对不对：I want to read more to develop my ability of concentration. 如果不对，哪里有问题？你将如何修改？为什么？

Electronic tools and resources are indispensible when translating texts, too. Try the following two short texts and see how electronic tools and resources can help with your translation. *(If you are too busy to translate them all, do finish the underlined parts. You could also try to do this translation with and without electronic tools/ resources and compare the effects.)*

11. 下面是一篇广东省应急办网站新闻，需要在半天内翻译成英文，因需要马上挂网，语言水平要求较高，需达到出版水平。翻译并记录下你的翻译过程。

甲型 H7N9 禽流感人类感染情况问答

H7N9 禽流感是一个全新的亚型，在全球尚属首次发现，该病种不属于原来的法定传染病。那么什么是 H7N9，怎么预防 H7N9，下面我们将详细解释。

问题 1：什么是 H7N9 型禽流感？
解答：H7N9 型禽流感是一种新型人感染禽流感，于 2013 年 3 月底在上海和安徽两地率先发现人感染疫情。感染 H7N9 型禽流感早期均表现出发热等症状，至目前尚未证实该病毒具有人际传播的可能性。禽流感是由流感病毒引起的感染性疾病，主要在禽类等动物之间流行，少数亚型可感染人。根据对人的致病力，禽流感病毒可分为高致病性禽流感病毒、低致病性禽流感病毒和无致病性禽流感病毒。
问题 2：H7N9 型禽流感是如何被命名的？
解答：H 和 N 分别代表什么含义？H 被称为红细胞凝聚素，N 被称为神经氨酸苷酶。这两者都是糖蛋白，分布在病毒表面。H 有 1～5 个亚型，N 有 1～9 个亚型。构成流感病毒遗传基因的核糖核酸的突然变异，比人体脱氧核糖核酸要快 100 万倍。由 H 和 N 的组合不同，病毒的毒性和传播速度也不相同。
问题 3：H7N9 型禽流感有哪些特点？
解答：H7N9 型禽流感是全球首次发现人感染该病毒，其生物学特点、致病力、传播力，还没有依据进行分析判断。

问题4：这次禽流感的感染来源是什么？

解答：既往国际上所发现的人感染 H7 亚型流感病毒多来自于禽类。但截至目前，国内所确诊的病例具体传播途径尚不清楚。自 1996 年到 2012 年，人类感染 H7 型流感病毒（H7N2、H7N3、H7N7）在荷兰、意大利、加拿大、美国、墨西哥以及英国都有病例报告。大多数感染与禽流感暴发相关。这些感染主要引发结膜炎以及轻度以上呼吸道症状，仅在荷兰发生过一例死亡。在此之前，中国从未有人感染过 H7 型流感病毒的报告。

问题5：H7N9 禽流感病毒的传染性是否严重？

解答：目前对 H7N9 型禽流感亚型及其感染力、致病力等的研究资料十分有限，尚无法对该病毒的毒力做出准确判断，尚未证实该亚型具有人传人的可能性。国内尚无针对性的疫苗。

2013 年 4 月 5 日，国家食品药品监督管理总局批准生产抗流感新药帕拉米韦氯化钠注射液，现有临床试验数据证明其对甲型和乙型流感有效。

问题6：人感染 H7N9 型禽流感有哪些症状？

解答：目前确诊病例主要表现为典型的病毒性肺炎，起病急，病程早期均可出现高热（38℃以上）、咳嗽等呼吸道感染症状。起病 5～7 天可出现呼吸困难、重症肺炎并进行性加重，部分病例可迅速发展为急性呼吸窘迫综合征并死亡。

问题7：所有的禽流感病毒都能引起人类的感染吗？

解答：答案是否定的。所有人类的流感病毒都可以引起禽类流感，但不是所有的禽流感病毒都可以引起人类流感。在禽流感病毒中，只有 H5、H7、H9 可以传染给人。从 1968 年到 1969 年之间，甲型 H3N2 病毒共造成了全球范围内 100 万人死亡。流感病毒有很严格的宿主特异性，禽流感病毒的基因组与人流感病毒（实际为能在人际传播的禽流感即 H1、H2 和 H3 型的基因组）有一些关键的差异，所以目前没有造成人与人之间的传播。

问题8：高致病性病毒的突变规律及危害如何？

解答：每隔 10～40 年就会出现一种高致病性病毒。

病毒平均每年有 10 次突然变异，30 年就有 300 次，大概在 300 次突然变异中，就有一次变成强毒型。在过去 100 年间，给人类带来巨大灾难的病毒有三种，分别是：导致西班牙流感的 H1N1 型，导致亚洲流感（1957 年）的 H2N2 型，以及导致香港流感（1968 年）的 H3N2 型。H5N1 型是堪与它们相提并论的强毒型病毒，值得予以警惕。

问题 9：H7N9 禽流感当前，我们还能吃禽肉与猪肉吗？

解答：烹调充分的食物不会传染流感病毒。因为流感病毒在通常的烹调温度下会失去活性，食物各部位均达 70℃，滚烫而没有红色带血部分之后再食用。经过正确加工与烹调的肉类是安全的，但不应食用患病及病死的禽类和畜肉，而食用生的以及未经充分烹调的带血肉类是高危行为，应该制止。

问题 10：个人如何避免感染甲型 H7N9 流感病毒？

解答：虽然目前国内外尚无针对 H7N9 禽流感病毒的疫苗，但是我们面对流感病毒也不用过分恐慌。只要通过不断学习，正确面对它，利用科学的方法还是可以有效的预防 H7N9 禽流感的。

开窗通风

无论是在办公室还是家里，都要注意加强室内空气流通，每天开窗 1～2 次，每次 30 分钟。

勤洗手

要勤洗手，无论是饭前便后，还是待人接物后，特别是在接触禽、蛋等东西后，一定要用消毒液、洗手液，五指交叉，活水洗手。

掩面而嚏

无论是在公共场所，还是在家中密闭空间，当要打喷嚏、咳嗽时，要用纸巾或手帕掩住口鼻，以免通过飞沫传播病菌。

戴口罩

如果要外出，去人员密集的公共场所，戴上口罩还是有一定的预防病毒传染的作用。

高温烹食

根据媒体宣传，H7N9 禽流感对外界环境抵抗力不强，100℃的高温一分钟便可消灭。因此，对煮食禽蛋类食品，一定要100℃的高温烹调数分钟后才能食用，不要吃半生不熟的鸡、鸭、鹅及鸡蛋等。

不接触禽类

如非必要，不要去花鸟园、养鸡场等有潜在禽类病毒传染源的地方，远离家禽的分泌物，更不要捕捉，喂食鸽子、野生鸟类等。

不去密闭场所

如非必要，不要去人员密集的密闭场所，如酒吧、夜店、电影院等人员密集、空气流通差的地方。

选购卫生食品

选购肉禽类食品，要到品质有保障的商店、超市购买有食品检验检疫机构检验合格的肉禽类产品，并做到高温煮食。

早睡早起

不熬夜，做到早睡早起，每晚11点前，务必入睡。通过高质量的睡眠，来提高自身的免疫力。

加强锻炼

多运动，可以到户外空气流通的地方，跑步、打羽毛球、打太极拳，通过锻炼来提高自身抵抗力。

补充营养

多吃蔬菜、水果，特别是富含维生素 C 的绿叶菜和水果，如苋菜、芹菜、西兰花、黄瓜、橙子、猕猴桃、小番茄等，通过摄取足够的营养来提高自身免疫力。

12. 下面是 *Understanding Philanthropy* 一书的前言开始几段，同上一题一样，因为译作需要出版，语言质量要求也较高。尝试翻译一下，记录下翻译过程，并对比一下翻译这篇文章和11 题的差异。

Preface and Acknowledgments

Unlike the rest of the book, where the voices of the two authors mix as one, in this preface we each contribute separately. This allows us to give a sense of how we each came to this book from our own perspective, background, and biases.

From Robert L. Payton

I've been writing professionally — that is, for publication — for more than fifty years. The book that follows focuses on philanthropy, one of the persistent themes that I've explored in that writing. A second fact of my professional life is that I've been a practitioner as well as a student and teacher of philanthropy. In my old age I've reflected on "experience" in every aspect of my life; I put a high value on experience as a test of my ideas and values. What I write about philanthropy is tested against my own personal experience as a practitioner of philanthropy and in light of the experience of others: employers, colleagues, students, volunteers, and my wife, who practices what I preach and tells me when practice and preaching conflict.

In addition to experience, my way of looking at philanthropy has been profoundly influenced by another fact of my professional life: I have spent several decades in colleges and universities, as administrator, editor, speech-writer, fund-raiser, teacher, and "scholar" — that is, lifelong student, not only of philanthropy but of many other things, with special interest in and emphasis on the humanities and liberal arts. Some years ago I discovered *the* idea of "the between," where the gods reach down to touch humanity and humanity reaches upward to touch divinity. "The gods" is a metaphor for knowledge; "humanity" is a claim of special status in human affairs for the search for truth, that search being the best of what makes us human.

Directions: After finishing doing the above activities, you'll have to share with the class your group's results (esp. how you have come to the results) and tools/resources you find helpful for you during translating. Pay attention, your presentation should cover at least *three of the six kinds* of must-use resources/tools in the tutor's recommendation.

Appendix 11   Sample Pages from the Brochure for Problem 3

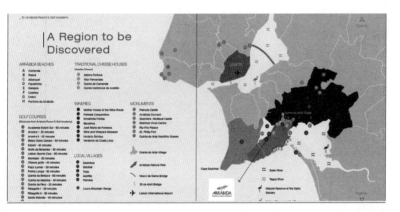

Appendix 12　Sample Pages from the Brochure for Problem 4

Using ApSIC Xbench

Version 3.0

Copyright Notice

ApSIC, S.L.
Caballero, 76 4-3
08029 Barcelona
Spain
Website: http://www.xbench.net

Contents

Overview

ApSIC Xbench allows you to organize and search your bilingual reference terminology. ApSIC Xbench also features several Quality Assurance (QA) checks to boost the quality of your translations.

Search features

ApSIC Xbench allows you to perform powerful searches on the following bilingual formats:

- Tab-delimited text files (*.txt, *.utx)
- XLIFF files, including MemoQ XLIFF, MemSource XLIFF, Idiom XLIFF and other flavors of XLIFF (*.xlf, *.xlif, *.xliff, *.xlz, *.mqxlz, *.mxliff)
- TMX memories (*.tmx)
- TBX/MARTIF glossaries (*.xml, *.tbx, *.mtf)
- TIFF files (*.tipp)
- Trados exported memories (*.txt)
- Trados exported MultiTerm 5 glossaries (*.txt)
- Trados MultiTerm XML glossaries (*.xml)
- Trados TagEditor files (*.ttx)
- Trados Word uncleaned files (*.doc, *.rtf)
- Trados Studio files (*.sdlxliff, *.sdlproj)
- Trados Studio memories (*.sdltm)
- memoQ files, including hand-off and hand-back packages (*.mqxlz, *.mqxliff, *.mqout, *.mqback)
- SDLX ITD files (*.itd). **Note:** This option requires that SDLX is installed on the machine.
- SDLX memories (*.mdb)
- STAR Transit 2.6/XV directory trees
- PO files (*.po)
- IBM TranslationManager exported dictionaries (*.sgm)
- IBM TranslationManager installed and exported folders (*.fxp)
- IBM TranslationManager exported memories (*.exp)
- OpenTM2 exported dictionaries (*.sgm)
- OpenTM2 installed and exported folders (*.fxp)
- OpenTM2 exported memories (*.exp)
- Wordfast memories (*.txt)

Appendix 13　The Fourth Problem（Full Version）

1. 项目作业说明
- 以小组协作形式完成 ApSIC Xbench 3.0 User Guide EN 的英译中翻译任务

- 字数：Word 统计 13, 033
- 页数：77 页
- 文档格式：PDF
- 项目完成时间：7 天
- 项目完成形式：小组合作 + 计算机辅助翻译工具，需完成从项目计划至项目总结整个流程

2. 项目总体要求
- 以小组协作形式完成《待翻文档 ApSIC Xbench 3.0 User Guide EN》的英译中翻译任务
- 完成从项目计划至项目总结整个流程
- 重复利用现有的 legacy（客户会提供）制作记忆库、术语库，以提高小组翻译的效率和质量
- 翻译过程中必须充分利用 Trados 2011 单机版的各项功能，项目总结报告中需提交软件使用评估报告
- 项目过程中可自行发掘其他在线或开源工具协助项目进行，如资源传输和共享工具等（鼓励在项目总结汇报时分享）

3. 工具使用要求
- 要利用老师提供的 2.9 版中英文文本制作记忆库、术语库，以提高小组翻译的效率和质量
- 翻译过程中必须充分利用 SDL Trados Studio 2011 单机版的各项功能，项目总结报告中需提交软件使用总结报告
- 项目过程中可自行发掘其他在线或开源工具协助项目进行，如资源传输和共享工具，如 Dropbox 等

4. 格式要求
- 最后提交译文 PDF 和 Word 文档
- 格式要保持与原文一致

- 译文中中文字体：宋体；字体大小保持与原文一致；译文中英文字体：Times New Roman
- 各部分文字颜色和 PDF 原文档颜色保持一致
- 原文文档中图片上文字不用翻译，但是图片需要保留在原有位置

5. 风格要求
- 本项目是软件使用手册，属于信息类文体，译文语言需要客观准确
- 中文表达要流畅、地道
- 原文中专有名词可保留原文也可以译成中文，请注意上下文一致即可
- 原文中的引号或加粗的强调方式，在译文中尽可能保留，实在无法保留的请自行根据中文习惯选择合适方式，注意上下文一致即可
- 译文术语、全文风格要有一致性

6. 最后提交文件要求
- 项目涉及的所有文件需要及时存档、妥善管理。最后提交的文件夹需要根据具体情况进行分类、管理，要求文件名规范、统一，文件夹结构清晰、合理，文件夹内容完整、一目了然
- 最后的"Final Delivery"文件夹里需要提交编辑好的译文、sdlxliff 双语文件（保留修订痕迹）、维护好的记忆库和术语库；其中记忆库所有针对此次翻译任务的部分需要添加字段：客户 — Xbench；领域 — 软件；术语库需要添加说明性字段：领域、定义、用法、例句、图片并各自定义好属性（各字段是否有内容，可以根据实际需要决定；可以留空）
- 最后呈交的文件夹内需包含但不限于下列内容：项目计划（初稿 & 终稿：项目简介、工作量分析、职责分工、流程、时间管理、风险管理、工具应用等），跟踪管理文件［译员培训文件、过程监督文件、项目成员交流记录、会议纪要、项目成员工作评估（自评 & 互评）用文件等］，翻译产品文件（详见第 2 点），质量管理文件（Style Guide、翻译质量评估表、翻译质量标准、QA 汇总、翻译错误案例集等），项目报告（项目管理总结 & 软件功能评估）

注：项目管理总结报告里无需重复计划里已经确定的内容，只需对项目实施进行经验教训的总结和反思即可，记住需包括对项目耗时的统计。

Appendix 14 Rubrics for Grading the Summative Assessment Tasks

Assessment tasks	ILOs	Percentage points	Scale score				
			A Excellent	B Good	C Adequate	D Marginal	F Failure
			5	4	3	2	1
Portfolio	ILO 1,4, 12	5	Showing great passion about and capability of learning and applying CAT with not only ample learning materials but also enough application cases	Showing enough passion about and capability of learning and applying CAT with ample learning materials and limited attempts at its application	Showing moderate self-initiated efforts to learn CAT with ample learning materials but without any efforts to apply it	Showing some self-initiated efforts to learn CAT with a few learning materials but without any efforts to apply it	Showing no self-initiated efforts to learn CAT with little learning materials and no efforts to apply it

Continued

Assessment tasks	ILOs	Percentage points	Scale score				
			A Excellent	B Good	C Adequate	D Marginal	F Failure
			9-10	7-8	5-6	3-4	1-2
Essay questions	ILOs 1, 2, 4, 6	10	Showing thorough understanding of the key concepts by explaining them articulately and logically with supporting details	Showing enough understanding of the key concepts by explaining them articulately and logically though without many desired details	Showing proper understanding of the key concepts by explaining them clearly without any details	Showing some understanding of the key concepts but explain barely clearly	Showing little understanding of the key concepts without any explanation

Continued

	Assessment tasks	ILOs	Percentage points	Scale score				
				A *Excellent*	B *Good*	C *Adequate*	D *Marginal*	F *Failure*
Problem 4	Plan		15					
	Work flow design	ILOs 6, 7, 9	7	7	5-6	3-4	1-2	0
				With perfectly clear, reasonable and feasible work flow design	With sufficiently clear, reasonable and feasible work flow design	With moderately clear, reasonable and feasible work flow design	With barely clear, reasonable and feasible work flow design	Without clear, reasonable and feasible work flow design
	Technological aids	ILOs 1, 2, 4, 7, 9	8	7-8	5-6	3-4	1-2	0
				Adopting a variety of CAT tools perfectly suitable for their purposes	Adopting a variety of CAT tools sufficiently suitable for their purposes	Adopting a variety of CAT tools moderately suitable for their purposes	Adopting a variety of CAT tools barely suitable for their purposes	Adopting no CAT tools or tools that are not suitable for their purposes

Continued

Assessment tasks			ILOs	Percentage points	Scale score				
					A Excellent 25-30	B Good 19-24	C Adequate 13-18	D Marginal 7-12	F Failure 1-6
Problem 4	Implementation	Process management	ILOs 3, 5, 6, 10	30 / 30	Showing great ability to organize and implement the translation project effectively with well-organized process documents, efficient communication and cooperation, and skillful risk control	Showing good ability to organize and implement the translation project effectively with well-organized process documents, efficient communication and cooperation, and skillful risk control	Showing moderate ability to organize and implement the translation project effectively with well-organized process documents, efficient communication and cooperation, and skillful risk control	Showing poor ability to organize and implement the translation project effectively with well-organized process documents, efficient communication and cooperation, and skillful risk control	Showing no ability to organize and implement the translation project effectively with well-organized process documents, efficient communication and cooperation, and skillful risk control

Continued

Assessment tasks		ILOs	Percentage points	Scale score				
				A Excellent	B Good	C Adequate	D Marginal	F Failure
				9-10	7-8	5-6	3-4	1-2
Problem 4	Implementation	Translation product	ILOs 3, 5, 10	10	Meeting perfectly the client's requirements in terms of accuracy, style, and format	Meeting sufficiently the client's requirements in terms of accuracy, style, and format	Meeting moderately the client's requirements in terms of accuracy, style, and format	Meeting poorly the client's requirements in terms of accuracy, style, and format	Meeting not the client's requirements in terms of accuracy, style, and format
	Reflection	Technological aids evaluation	ILOs 1, 2, 4, 5, 7, 8	10					
				9-10	7-8	5-6	3-4	1-2
				Showing perfectly critical and well-justified evaluation of all the tools adopted	Showing sufficiently critical and well-justified evaluation of all the tools adopted	Showing moderately critical and justified evaluation of all the tools adopted with	Showing poorly critical evaluation of all the tools adopted, without justification	Showing no critical and well-justified evaluation of all the tools adopted

Continued

Assessment tasks			ILOs	Percentage points	Scale score				
					A Excellent	B Good	C Adequate	D Marginal	F Failure
					5	4	3	2	1
Problem 4	Reflection	Reflective report	ILOs 1, 2, 4-7, 9-12	5	Showing sound knowledge of and reflections on the translation project management and great ability to assess objectively the performance of one's own and his/her partners	Showing sufficient knowledge of and reflections on the translation project management and good ability to assess objectively the performance of one's own and his/her partners	Showing moderate knowledge of and reflections on the translation project management and mediocre ability to assess objectively the performance of one's own and his/her partners	Showing scant knowledge of and reflections on the translation project management and poor ability to assess objectively the performance of one's own and his/her partners	Showing little knowledge of and reflections on the translation project management and no ability to assess objectively the performance of one's own and his/her partners

Continued

Assessment tasks		ILOs	Percentage points	Scale score				
				A Excellent	B Good	C Adequate	D Marginal	F Failure
				5	4	3	2	1
Problem 4	Reflec-tion / Experience exchange	ILOs 1, 2, 4-9, 12	5 / 5	Expressing the experience and lessons obtained during the project in a perfectly concise, reflective and articulate way	Expressing the experience and lessons obtained during the project in a sufficiently concise, reflective and articulate way	Expressing the experience and lessons obtained during the project in a moderately concise, reflective and articulate way	Expressing the experience and lessons obtained during the project in a marginally concise, reflective and articulate way	Expressing the experience and lessons obtained during the project in a barely concise, reflective and articulate way

Appendix 15　Rubrics for Assessing the ILOs as Evidenced by the Assessment Tasks

ILOs	Content	Assessment tasks	Percentage points	Scale score				
				A Excellent 7	B Good 5-6	C Adequate 3-4	D Marginal 1-2	F Failure 0
ILO 1	Be able to define CAT and MT, name a few leading systems of each and describe their mechanics, functions, development and applications against the background of the translation profession	Portfolios & Essay questions & Problem 4 (Plan-Technological aids; Implementation-Process management; Reflection)	7	Showing thorough understanding of the key concepts	Showing sufficient understanding of the key concepts	Showing enough understanding of the key concepts	Showing poor understanding of the key concepts	Showing no understanding of the key concepts

Continued

ILOs	Content	Assessment tasks	Percentage points	Scale score					
				A *Excellent*	B *Good*	C *Adequate*	D *Marginal*	F *Failure*	
				3	2		1	0	
ILO 2	Be able to explain the differences between CAT and MT	Essay questions & Problem 4 (Plan-Technological aids; Implementation-Process management; Reflection)	3	Knowing clearly the differences and displaying a systematic comparison between them	Knowing some of the differences and displaying a comprehensive between them	logical if not comparison	Knowing some of the differences but unable to make reasonable comparison	Knowing nothing about the differences	

Continued

ILOs	Content	Assessment tasks	Percentage points	Scale score				
				A Excellent 16-18	B Good 11-15	C Adequate 6-10	D Marginal 1-5	F Failure 0
ILO 3	Be able to operate skillfully at least one narrow-sense CAT system among SDL Trados 2011, Oriental Yaxin and Google Translator Toolkit, to the effect that it can assist with his/her translation practice effectively	Problem 4 (Implementation)	25	Operating very skillfully one of the specified systems, knowing perfectly how, when and for what to use it	Operating skillfully one of the specified systems, knowing sufficiently how, when and for what to use it	Operating with certain confidence one of the specified systems, knowing moderately how, when and for what to use it	Operating poorly one of the specified systems, not knowing exactly how, when and for what to use it	Unable to operate any of the specified systems

ILOs	Content	Assessment tasks	Percentage points	Scale score				
				A Excellent	B Good	C Adequate	D Marginal	F Failure
				9-10	6-8	3-5	2-4	0-1
ILO 4	Be able to find and classify broad-sense CAT tools available to the translator;	Portfolios & Essay questions &Problem 4 (Plan-Technological aids; Reflection)	10	Knowing perfectly a wide array of broad-sense CAT tools and their functions	Knowing sufficiently a wide array of broad-sense CAT tools and their functions	Knowing moderately a wide array of broad-sense CAT tools and their functions	Knowing poorly broad-sense CAT tools and their functions	Knowing not broad-sense CAT tools
ILO 5	Be able to apply skillfully those broad-sense CAT tools covered in this course in translation learning and practice;	Problem 4 (Implementation; Reflection)	10	9-10	6-8	3-5	2-4	0-1
				Applying very skillfully those broad-sense CAT tools covered in this course in translation learning and practice	Applying skillfully those broad-sense CAT tools covered in this course in translation learning and practice	Applying moderately skillfully those broad-sense CAT tools covered in this course in translation learning and practice	Applying poorly those broad-sense CAT tools covered in this course in translation learning and practice	Applying no broad-sense CAT tools covered in this course in translation learning and practice

Continued

ILOs	Content	Assessment tasks	Percentage points	Scale score				
				A Excellent	B Good	C Adequate	D Marginal	F Failure
ILO 6	Be able to explain what translation project is and how it is normally managed;	Essay questions & Problem 4 (Plan-Work flow design; Implementation-Process management; Reflection-Reflective report, Experience exchange)	5	5 Be able to explain articulately and logically what translation project is and how it is normally managed	4 Be able to explain quite clearly translation project is and how it is normally managed	3 Be able to explain moderately clearly what translation project is and how it is normally managed	1-2 Be able to explain but rather poorly what translation project is and how it is normally managed	0 Be unable to explain articulately and logically what translation project is and how it is normally
ILO 7	Be able to analyse a given translation task/problem with a view to selecting appropriate methods/tools for it;	Problem 4 (Plan; Reflection)	10	9-10 Being able to select various tools that fit perfectly with a translation task	6-8 Being able to select various tools that fit sufficiently with a translation task	3-5 Being able to select various tools that fit moderately with a translation task	2-4 Being able to select some tools suitable a translation task	0-1 Being unable to select various tools that fit with a translation task

Continued

ILOs	Content	Assessment tasks	Percentage points	Scale score				
				A Excellent	B Good	C Adequate	D Marginal	F Failure
				9-10	6-8	3-5	2-4	0-1
ILO 8	Be able to evaluate critically the usability and efficacy of adopted tools and resources and make well-grounded judgment about their values and suitability in different working contexts;	Problem 4 (Reflection-Technological aids evaluation, Experience exchange)	10	Being able to evaluate critically the usability and efficacy of adopted tools and resources and make well-grounded judgment about their values and suitability in different working contexts	Be able to evaluate moderately critically the usability and efficacy of adopted tools and resources and make judgment about their values and suitability in different working contexts though not justifiably	Be able to evaluate the usability and efficacy of adopted tools and resources and make judgment about their values and suitability in different working contexts, but without much justification	Be aware of the need to evaluate the usability and efficacy of adopted tools and resources and make their judgment about values and suitability in different working contexts but unable to carry them out	Be unable to evaluate critically the usability and efficacy of adopted tools and resources and make well-grounded judgment about their values and suitability in different working contexts

Continued

ILOs	Content	Assessment tasks	Percentage points	Scale score				
				A Excellent	B Good	C Adequate	D Marginal	F Failure
ILO 9	Be able to plan a technologically assisted translation project as needed and reflect on its effects at its conclusion;	Problem 4 (Plan; Reflection-Reflective report; Experience exchange)	6	6 — Making a highly feasible and reasonable plan of a translation project and well-justifiably evaluation of its effectiveness upon reflection	4-5 — Making feasible and reasonable plan of a translation project and moderately well-justifiably evaluation of its effectiveness upon reflection	3 — Making a feasible though not reasonable enough plan of a translation project and evaluating its effectiveness moderately justifiably	2 — Making a poor plan of a translation project and judging its effectiveness rather subjectively	1 — Making a very poor plan of a translation project and judging its effectiveness entirely subjectively
ILO 10	Be able to communicate and cooperate with peers efficiently to achieve a common goal;	Problem 4 (Implementation)	4	4 — Communicating and cooperating with peers efficiently to successfully achieve a common goal	3 — Communicating and cooperating with peers moderately efficiently to achieve a common goal	2 — Communicating and cooperating with peers only satisfactorily to achieve a common goal	1 — Communicating and cooperating with peers poorly to successfully achieve a common goal	0 — Failing to communicate and cooperate with peers efficiently to achieve a common goal

Continued

ILOs	Content	Assessment tasks	Percentage points	Scale score				
				A Excellent	B Good	C Adequate	D Marginal	F Failure
ILO 11	Be able to monitor and assess the performance reasonably during solving a translation problem both individually and collectively;	Problem 4 (Implementation; Reflection-Reflective report)	4	4 Monitoring and assessing the performance highly reasonably during solving a translation problem both individually and collectively	3 Monitoring and assessing the performance reasonably during solving a translation problem both individually and collectively	2 Monitoring and assessing the performance moderately reasonably during solving a translation problem both individually and collectively	1 Monitoring and assessing the performance poorly during solving a translation problem both individually and collectively	0 Unable to monitor and assess the performance highly reasonably during solving a translation problem both individually and collectively
ILO12	Be able to learn CAT proactively and continuously.	Portfolio & Problem 4 (Reflection-Reflective report)	6	6 Showing great ability to learn CAT proactively and continuously	4-5 Showing sufficient ability to learn CAT proactively and continuously	2-3 Showing mediocre ability to learn CAT proactively and continuously	1 Showing poor ability to learn CAT proactively and continuously	0 Showing no ability to learn CAT proactively and continuously

Appendix 16　Essay Questions

各位同学：

　　大家好！下面的题目均来自平时给大家的思考题，用于考察大家对 CAT 基本知识的掌握程度。每人需选择回答至少 5 道题，其中第 1、6 和 12 题为必答题，在剩余的题目中再选 2 题回答即可。

　　再次感谢你们的辛苦努力和配合，让我们一起度过了这个愉快而难忘的旅程！

1. 机器翻译和计算机辅助翻译有什么区别？［必答题］
2. 请举例说明如何利用 Google 高级检索语法辅助翻译工作。
3. 请简单说明文学文本和非文学文本的差异，以及文本类型和 CAT 工具选择有什么关系。
4. 什么是术语？术语对技术写作有什么作用？
5. Match 在专业 CAT 工具中对应的中文是什么？Trados 中的 match 有几个层次？分别是什么意思？
6. 在翻译项目的初期为什么需要应用 WinAlign、Multiterm Extract 或雅信的专家库建设和检索平台？它们的主要功能是什么？［必答题］
7. 什么是 Controlled Language（CL）？它有哪些基本规则？为什么很多公司要采用 CL 进行写作？
8. 在 SDL Trados Studio 上翻译审校完成后，是否可以进行自动检验？对于检验内容是否可以自己进行定义？在哪里进行？
9. 请列举至少 5 个 Trados Studio 上常用的快捷键。
10. Trados 的主记忆库和项目记忆库有何区别？它们如何更新？
11. Trados 的 WinAlign 中，什么是标准连接，什么是强制 1∶1 连接？
12. 请简述翻译项目管理的 7 个步骤，并说明在课程最后一个翻译项目中你体会最深的是什么。［必答题］

Appendix 17　A Sample Individual Journal from the Learners

Student Name:　洪 ＊　Date:　2013-08-22
Task Description:【关于"我只是占了时代的便宜"】
Time Span:【21: 00 ～ 22: 00】
Tools/Purpose: 关于李继宏重译《瓦尔登湖》
Resources/Purposes:【ps. 这是一篇"观"后感】

最近我在微博上关注的几个翻译大家都在评论这件事情。今天下午，终于看到了这篇演讲稿。

说实话，我并没有觉得太过愤怒。尽管关注的大咖们纷纷斥之为"没有底线的无耻营销"，认为他"为了给自己打广告妄图踩断前人的肩膀"；尽管我自己清楚他的话与事实有悖，并且对他的一些观点表示全然反对。但是他却说了不知是有心还是无心的"真理"："我只是占了时代的便宜。"

在这点上，他说对了。就像他自己举的例子："Innocent X"不是"天真的 X"而是"因诺森特十世"，"The Fifth Avenue"不是"五号街"而是"第五大道"……在那个年代，真正出过国的翻译家又能有几个？他们手上真正准确权威的词典，又能有几本？哪像现在：即使是一个拜托你帮忙翻译的连英文字母有几个都数不清的亲戚，都会在后面添上一句："不要偷懒用 Google Translator 哦 ～"

网络、共享理念、语料库、CAT……这些都是时代发展带来的改变。

李继宏自己也承认：《瓦尔登湖》翻译的一大难点，就是数之不尽的动植物名称。但这些"术语"，只要找对了工具，要翻译核实并不会太困难。当然，从这个意义上讲，他自称的这本"最好的翻译版本"，倒也是对的。

只是，他忘了一点：好的翻译家，除了现在必须是个 IT 达人，本质上还是个"码字的"。

冰心、杨绛、傅雷、余光中、施咸荣……他们的译本完美吗？当然不完美。可是我看他们的译本，觉得是在看"一部作品"。

而李继宏、林淑娟、贾秀琰……他们的译本科学吗？科学。可是我看他们的译本，觉得是在看"一个译本"。

我虽浅薄，却也知道一个字的斟酌有多不易：这是我内在修行不够。而看冰心老人他们的译作，洋洋洒洒，字字珠玑；纵然有着现在看来"明显的错误"，却毫不妨碍我去沉浸在他们的世界。在我看来，一本译作，虽不能是译者的想当然的"再创作"，却可以是译者思考后的一种"修缮"。这就像我当时跟朋友评论冯小刚的那部电影《一九四二》时说的一样："这是一部完整的作品，但我却看不到冯小刚的思考。如果我仅仅想知道这段历史，我完全可以找我的历史老师，何必花费这 60 块钱买一堆烂砖瓦。"

但这样说来，我反倒有些想看看李继宏这一版的《瓦尔登湖》了。我自傲自己小学时就已阅过无数外国著作的经典译本，却唯独没能安安静静看完《瓦尔登湖》。纵然他没能翻译出我心目中的《追风筝的人》，倒也不妨碍我去看看他翻的另一种风格的文字。只是这次，他碰到的是公认的经典，而不再是一本现代的看过就可丢弃的畅销小说。

Results and Reflection:
Merits:【无】
Demerits:【其实有些比较偏激的话都删掉了，总感觉自己是在以己之意度他人之心了。实话实说，我很佩服李继宏老师，他的经历与他的思考，都有可圈可点的地方。而且难能可贵的是，他承认他的私心，并为之做了有理的辩护，很佩服这样自信的人。】
Daily Self-assessment: (4) points
【记在后边儿 ~搜索过程中一些有意思的文档】
这篇是原文：http://weibo. com/1182419921/A5TJp9pcV。